PRIMER FOR LINEAR ALGEBRA

Stephen Demko
Georgia Institute of Technology

Scott, Foresman and Company
Glenview, Illinois London, England

Printed in the United States of America.

ISBN-0-673-38642-2

1 2 3 4 5 6 - MAL - 93 92 91 90 89 88

PREFACE

This book is meant to provide an introduction to some of the important aspects of linear algebra for engineering and science students in their sophomore year. The contents and emphasis reflect my belief that the first exposure to linear algebra have a strong geometric emphasis and be broadly based. As such the emphasis in Chapters 3 through 5, the core of the book, is on two and three dimensional Euclidean spaces and a variety of topics are discussed: least squares problems, matrix representations of rigid motions, eigenanalysis and applications. Detailed treatments of the infrastructure of linear algebra – subspaces, general vector spaces, bases, complex vector spaces, etc. – must be left for a future course.

The basis for this book is a set of notes I developed over a period of years as part of a multivariate calculus course. I typically presented selected topics from Chapters 3 through 5 together with Appendices A through C. Elements of Chapters 1 and 2 are introduced as needed, although most students know some method for solving linear systems by this time. Here are some of the more distinctive features of this book.

In Section 1.3 a method of saving multipliers used in Gaussian Elimination is presented. This is essentially the LU-factorization without matrices.

In Section 2.3 the inverse is presented as a useful concept, but its use in the solution of linear systems is discouraged.

In Sections 3.2 and 3.5 overdetermined and underdetermined linear systems are discussed. The solution of the former makes use of the concept of column space. In the development of the solution of underdetermined systems the null space plays a crucial role. I think that it is desirable to cover one of these sections to keep the course concrete.

Chapter 4 is devoted to representations of affine and linear functions, mostly in the plane. The emphasis is on finding algebraic expressions for geometrically defined functions: rotations, reflections, and projections.

Chapter 5 provides a geometric introduction to eigenanalysis and a few of the standard applications. Complex eigenvalues are not discussed but their existence is noted.

The Appendices A through C contain material that is nice to have in multivariate calculus courses but does not appear in every book. The Newton's method exercises provide a superficial introduction to Fractal Geometry.

Appendix D is recreational in nature. It is a nice example of the utility of affine maps.

An Answers/Solutions manual is available from the publisher.

TABLE OF CONTENTS

CHAPTER 1 The Solution of Systems of Linear Equations

This chapter contains a discussion of systems of linear equations and their solution. An efficient computational technique called Gaussian Elimination is developed. The goal of this chapter is to provide a reliable, robust method for solving systems of linear equations which the reader can implement either by hand or on a computer or calculator.

1.1. Introduction to Linear Systems

We first discuss the solution of one linear equation in one unknown. The general equation we mean is

$$ax = b$$

where a and b are given numbers. You may think there's not much to talk about — the answer is obviously $x = \frac{b}{a}$. This is almost true. This is the one and only answer provided that a is not zero.

If $a = 0$, we have the equation

$$0x = b.$$

Your first inclination is to say that there is no solution. Again, this is almost true. If b is not zero, then there is *no* solution. However, if $b = 0$, then for any x the equation $0x = 0$ is true. So, there are *infinitely* many solutions in this case.

A *linear equation* in the two variables x and y is an equation of the form

$$ax + by = c \tag{1}$$

where a, b, and c are given numbers. The set of (x,y) values that make such an equation true determines a straight line in the Cartesian plane unless a and b are both zero while c is not, in which case there is no solution. This is why this kind of equation has the name "linear". In three variables the general form of a single linear equation is

$$ax + by + cz = d \tag{2}$$

where a, b, c, and d are assumed to be given. The solution set, i.e., the set of (x,y,z) values that make the equation true, is a plane in three-dimensional space whose normal vector is (a,b,c) but the equation is still called "linear".

In general we have the

Definition. A *linear equation* in the n-variables $x_1,x_2,\ldots,x_n$ is an equation of the form

$$a_1x_1 + a_2x_2 + \ldots + a_nx_n = b \tag{3}$$

where $a_1,a_2,\ldots,a_n$, and b are constants.

In the general case there is no easily visualized geometric entity associated with the set of solutions; but it is natural to use geometric terms to describe features of these equations. For example, the solution set of equation (3) is called a *hyperplane in n-dimensional Euclidean space.*

We say that we have a *system of linear equations* if we are faced with solving several linear equations simultaneously. For example, consider the problem of finding x, y, and z that solve the following system

$$\begin{aligned} 2x + 4y + z &= 0 \\ x - 2y + z &= 2. \end{aligned}$$

You might be able to find some solutions by inspection or even find an expression for all solutions. Shortly, we'll discuss a *systematic* method for finding all solutions to this and more general linear systems. It is worth pointing out that in the above system each equation is the equation of a plane. Since these planes have nonparallel normal vectors, they must intersect in a line. So, even before we do a computation we know the correct form of the complete solution set: there are infinitely many solutions and they form a line in three-dimensional space. Algebraically, this means that the solution must be expressible in terms of the parametric equations of a line. If these equations arose in some scientific context, then it might also be possible to ascribe some appropriate scientific meaning to the solution set. These algebraic, geometric, and scientific attributes of the solution set of any kind of equation are important — when they exist — if only because they allow a quick plausibility check of the computed answer.

Triangular Systems

Systems that have the form

$$\begin{aligned} ax + by + cz &= d \\ ey + fz &= g \\ hz &= k \end{aligned} \tag{4}$$

can be easily solved by solving the last equation and using the z value found there to reduce the second equation to one involving only y. The first equation then contains only one unknown, x, and can be easily solved. If this process breaks down, for example, if $h = 0$, then there is either no solution or infinitely many solutions. In the next section we will see how to transform a general system to one equivalent to (4). Right now we will illustrate what can happen in systems like (4). We call such systems *upper triangular* because of the pattern of the unknowns.

Example 1. Solve the system

$$\begin{aligned} 2x + 3y \quad &= 4 \\ y + z &= 3 \\ 2z &= 6. \end{aligned}$$

Solution. From the last equation, $z = 3$. Substituting this in the second equation gives $y + 3 = 3$, so $y = 0$. Finally, setting $y = 0$ in the first equation gives $2x + 0 = 4$, so $x = 2$. The solution is $x = 2$, $y = 0$, $z = 3$. □

Example 2. Solve the system

$$\begin{aligned} x - 2y + 3z &= 4 \\ y + 5z &= 8 \\ z &= 9. \end{aligned}$$

Solution. $z = 9$ is given in the last equation. Setting $z = 9$ in the second equation, we get $y + 45 = 8$ so $y = -37$. Finally, with $z = 9$ and $y = -37$ in the first equation we have $x - 2(-37) + 3(9) = 4$ or $x = -97$. So, the solution is $x = -97$, $y = -37$, $z = 9$. □

The solution set can be infinite as the next example shows.

Example 3. Find the solution set of the system

$$\begin{aligned} 2x - y + 4z &= 8 \\ y - 2z &= 4. \end{aligned}$$

Solution. There are fewer equations than unknowns. Geometrically, the solution set is the set of all points on the intersection of the planes determined by the equations. Since the intersection of two planes is a line, we expect to have infinitely many solutions. From the second equation, we have $y = 2z + 4$. Substituting this in the first equation we get $2x - (2z+4) + 4z = 8$ or $2x + 2z = 12$. Therefore, $x = -z + 6$. So, the solution set can be expressed as follows: $z =$ anything, $y = 2z + 4$, $x = -z + 6$. If we think of the value of z as being given by a parameter t, we write $x = -t + 6$, $y = 2t + 4$, $z = t$. □

The basic technique applies to systems of any size as long as the equations have the correct structure. Here is an example with four variables.

Example 4. Find the solution set of the system

$$\begin{aligned} x_1 - 4x_2 + 5x_3 \qquad &= 0 \\ 2x_2 + x_3 + 3x_4 &= 8 \\ 2x_3 + 6x_4 &= 20 \\ 3x_4 &= 12. \end{aligned}$$

Solution. From the last equation $x_4 = 4$. This makes the third equation be $2x_3 + 24 = 20$, so $x_3 = -2$. The second now becomes $2x_2 - 2 + 12 = 8$, so $x_2 = -1$. Finally, the first equation reduces to $x_1 - 4(-1) + 5(-2) = 0$, so $x_1 = 6$. In summary, the solution is $x_1 = 6$, $x_2 = -1$, $x_3 = -2$, $x_4 = 4$. □

There may be no solutions as the next example demonstrates.

Example 5. Find the solution set of the system

$$\begin{aligned} 2x_1 - x_2 + 4x_3 + 8x_4 &= 10 \\ 2x_3 + 8x_4 &= 16 \\ x_3 + 4x_4 &= 9 \\ 3x_4 &= 3. \end{aligned}$$

Solution. From $x_4 = 1$ the third equation becomes $x_3 + 4 = 9$ and the second reduces to $2x_3 + 8 = 16$. However, these equations are incompatible since in the first case we get $x_3 = 5$ and in the second $x_3 = 4$. Therefore, there are no solutions. □

Here is one more system with infinitely many solutions.

Example 6. Find all solutions of the system

$$\begin{aligned} 2x_1 + x_2 + x_3 - x_4 &= 0 \\ 2x_3 + x_4 &= 8 \\ 4x_3 + 2x_4 &= 16 \\ x_4 &= -2. \end{aligned}$$

Solution. With $x_4 = -2$, the third equation becomes $4x_3 = 20$ or $x_3 = 5$. The second equation becomes $2x_3 = 10$ or $x_3 = 5$. This *is* compatible with the third equation. Using $x_4 = -2$ and $x_3 = 5$ in the first equation, we get $2x_1 + x_2 + 5 + 2 = 0$ or $2x_1 + x_2 = -7$. We have no additional equations to further specify x_1 or x_2 so we will simply let one of them be given by the parameter t and express the other in terms of t. So, the solution can be written as $x_1 = -\frac{1}{2}\,t - \frac{7}{2}$, $x_2 = t$, $x_3 = 5$, and $x_4 = -2$. Another way to write the solution is $x_1 = t$, $x_2 = -2t - 7$, $x_3 = 5$, $x_4 = -2$. In either case there is one free parameter. In some applications it would be useful to think of this solution set as being a line in a four-dimensional space. □

In our final example, the system is not "triangular" but it can be rearranged to be so.

Example 7. Solve the system

$$\begin{aligned} 2x_2 - x_3 &= 10 \\ x_1 + x_2 + x_3 - x_4 &= 0 \\ 3x_3 + x_4 &= 16 \\ x_4 &= 10. \end{aligned}$$

Solution. Interchange the first two equations to get

$$\begin{aligned} x_1 + x_2 + x_3 - x_4 &= 0 \\ 2x_2 - x_3 &= 10 \\ 3x_3 + x_4 &= 16 \\ x_4 &= 10. \end{aligned}$$

Then, $x_4 = 10$ and $3x_3 + 10 = 16$ gives $x_3 = 2$. The second equation becomes $2x_2 - 2 = 10$ so $x_2 = 6$. Finally, $x_1 = x_4 - x_3 - x_2 = 10 - 2 - 6 = 2$. □

1.1. EXERCISES

1. Identify the linear equations.

(a) $4x + 3y + z = 4$ (d) $2z = 1$ (g) $\sin x + y = 2$

(b) $x^2 + x = 2$ (e) $3x + 4y - 3z + w = 0$ (h) $x = y$

(c) $xy = 0$ (f) $e^2x + y = 0$ (i) $y^2 = x$

2. Each of the following linear systems can be thought of as representing two straight lines in the plane. Use what you know about lines to determine whether or not the system has exactly one solution, no solution or infinitely many solutions.

(a) $\begin{aligned} x - y &= 2 \\ x + y &= 4 \end{aligned}$ (c) $\begin{aligned} 4x - 3y &= 10 \\ 8x - 6y &= 21 \end{aligned}$ (e) $\begin{aligned} 3x &= 4 \\ x + y &= 0 \end{aligned}$

(b) $\begin{aligned} x + y &= 0 \\ x + 2y &= 1 \end{aligned}$ (d) $\begin{aligned} 4x + 5y &= 6 \\ 8x + 10y &= 12 \end{aligned}$ (f) $\begin{aligned} 2x - 2y &= 0 \\ 4x - 2y &= 0. \end{aligned}$

3. Each of the following upper triangular systems has a unique solution. Find it.

(a)
$$\begin{aligned} 2x - 3y + z &= 10 \\ y + z &= 2 \\ z &= 4 \end{aligned}$$

(b)
$$\begin{aligned} 4x - y &= 2 \\ 3y &= 8 \end{aligned}$$

(c)
$$\begin{aligned} x_1 - x_2 + x_3 - x_4 &= 0 \\ x_2 + x_3 + x_4 &= 0 \\ x_3 - x_4 &= 0 \\ x_4 &= 2 \end{aligned}$$

(d)
$$\begin{aligned} x - 2y + w &= 2 \\ 4y + 3z + 2w &= 2 \\ 4z - w &= -1 \\ w &= 1 \end{aligned}$$

(e)
$$\begin{aligned} x + y + z + w + u &= 10 \\ y + z + w + u &= 9 \\ 2z + w &= 10 \\ w - u &= 4 \\ u &= 0 \end{aligned}$$

4. Find the solution sets of the following systems

(a)
$$\begin{aligned} x + y + z + w &= 0 \\ y + z + w &= 4 \\ z + w &= 1 \end{aligned}$$

(b)
$$\begin{aligned} 2x - y + z &= 3 \\ x - y &= 4 \\ x &= 1 \end{aligned}$$

(c)
$$\begin{aligned} 2x_1 + x_2 + 3x_3 + x_5 &= 1 \\ x_3 + 2x_4 + x_5 &= 3 \\ x_3 + x_5 &= 1 \\ x_4 - x_5 &= 0 \\ x_5 &= 2 \end{aligned}$$

(d)
$$\begin{aligned} 2x_1 + 3x_2 + x_3 - x_4 &= 0 \\ 4x_2 + x_3 + 2x_4 &= 1 \\ x_3 - 2x_4 &= 2 \\ 2x_3 - 4x_4 &= 8 \end{aligned}$$

5. For what value(s) of the parameter p will the system below have a solution?

$$\begin{aligned} 2x_1 + x_2 - x_3 &= 0 \\ x_3 - x_4 &= p \\ 2x_3 + x_4 &= 1 \\ x_4 &= 3 \end{aligned}$$

1.2. Gaussian Elimination

We now discuss a method for solving general linear systems. The method, called *Gaussian Elimination*, transforms a general system into one which has the same solution set but which is upper triangular. As we've seen in the previous section, upper triangular systems are easy to solve.

We will use the following three operations to carry out the transformation described above:

Op. 1. Multiply an equation by a nonzero constant.

Op. 2. Subtract one equation from another and keep the resulting equation and one of the original two.

Op. 3. Interchange the order of the equations.

It is clear that neither Op. 1 nor Op. 3 change the solution set of a system. It is not hard to show that Op. 2 also leaves the solution set unaltered.

The real workhorse of the method is a combination of Op. 1 and Op. 2. We will typically multiply one equation by a well-chosen number and then subtract the resulting equation from one of the other equations. We will only occasionally use Op. 3. The following two examples each show a typical step of Gaussian Elimination.

Example 1. Use a combination of Op. 1 and Op. 2 to eliminate the x term from the second equation in the system

$$\begin{aligned} 2x - 3y &= 4 \\ x + y &= 7. \end{aligned}$$

Solution. If we multiply the first equation by $\frac{1}{2}$, we get

$$x - \tfrac{3}{2}y = 2.$$

Now, we will replace the second equation by the result of subtracting this equation from the second and get the new second equation

$$\tfrac{5}{2}y = 5.$$

The equivalent system consists of the original first equation and the new second equation

$$\begin{aligned} 2x - 3y &= 4 \\ \tfrac{5}{2}y &= 5. \end{aligned}$$ □

In the above example, the fact that each equation had two unknowns was not used in selecting the multiplier that helped us eliminate x in the second equation. All that mattered was the ratio of the coefficients of x.

Example 2. Eliminate x_1 from the second equation in the system

$$\begin{aligned} 3x_1 - x_2 + 8x_3 + x_4 &= 10 \\ 2x_1 + x_2 + 5x_4 &= 1. \end{aligned}$$

Solution. If we multiply the first equation by $\frac{2}{3}$, we get the equation

$$2x_1 - \tfrac{2}{3}\, x_2 + \tfrac{16}{3}\, x_3 + \tfrac{2}{3}\, x_4 = \tfrac{20}{3}$$

Now, subtracting this equation from the second gives the system

$$\begin{aligned} 3x_1 - x_2 + 8x_3 + x_4 &= \tfrac{20}{3} \\ \tfrac{5}{3}x_2 - \tfrac{16}{3}x_3 + \tfrac{13}{3}x_4 &= -\tfrac{17}{3}. \end{aligned}$$ □

The complete Gaussian Elimination process consists of *repeatedly* applying the step illustrated in the above examples in an *orderly* fashion. Let the unknowns be $x_1, x_2, \ldots, x_n$. First, the variable x_1 is eliminated from all but one equation. The equation from which x_1 is not eliminated is often called the *pivot* equation for x_1. A multiple of this equation will be subtracted from every other equation that contains x_1. In Examples 1 and 2 the first equation was the pivot equation for the first unknown. After eliminating x_1 from all equations except one, we attack the subsystem consisting of all the equations from which x_1 has been eliminated. We select one of these equations to be the pivot equation for x_2 and use it to eliminate the x_2 variable from all of the equations in the subsystem. We next consider the subsystem consisting of all equations except the pivot equations for x_1 and x_2. In our subsystem we should have only the variables $x_3,\ldots,x_n$. We then continue this process of elimination until we run out of either equations or unknowns.

After the elimination phase we should have an upper triangular system to solve. This system may have exactly one solution, no solutions, or infinitely many solutions. The process of solving this upper triangular system is called *backsolving.* The following examples illustrate the complete method.

Example 3. Find all solutions of

$$\begin{aligned} x + y - 2z &= 4 \\ x + 3y - z &= 3 \\ x + 3y + z &= 5. \end{aligned}$$

Solution. We will eliminate x from the second and third equations by subtracting the first equation from each of them. The new system is

$$\begin{aligned} x + y - 2z &= 4 \\ 2y + z &= -1 \\ 2y + 3z &= 1. \end{aligned}$$

We now concentrate on the last two equations. We eliminate y from the third equation by subtracting the second equation from the third. This gives $2z = 2$. So, our system is

$$\begin{aligned} x + y - 2z &= 4 \\ 2y + z &= -1 \\ 2z &= 2. \end{aligned}$$

And the solution is easy: from the last equation we get $z = 1$; using this in the second equation, we get $2y + 1 = -1$ or $y = -1$. Finally, the first equation is now $x-1-2 = 4$ so $x = 7$. You can check that the values $x = 7$, $y = -1$, $z = 1$ satisfy all of the original equations. □

Example 4. Find all solutions of the system

$$\begin{aligned} x - 2y + z &= 3 \\ 2x + 2y + 6z &= 14 \\ x + y + 3z &= 12. \end{aligned}$$

Solution. We first eliminate x from the second and third equations. Multiplying the first equation by 2 and subtracting the result from the second gives the new second equation $6y + 4z = 8$. Subtracting the first equation from the third gives the new third equation $3y + 2z = 9$. So, we now have the system

$$\begin{aligned} x - 2y + z &= 3 \\ 6y + 4z &= 8 \\ 3y + 2z &= 9. \end{aligned} \tag{1}$$

Now, to eliminate y from the third equation, we should multiply the second equation by $\frac{1}{2}$ and subtract the resulting equation from the third equation. This gives the system

$$\begin{aligned} x - 2y + z &= 3 \\ 6y + 4z &= 8 \\ 0y + 0z &= 5. \end{aligned}$$

The last equation has no solutions of course. Consequently, there are no solutions to the original system. You might have recognized that, after x was eliminated from the second and third equations, the second and third equations in (1) took the form of two parallel lines in the y-z plane, namely $6y + 4z = 8$ and $3y + 2z = 9$. This already showed there was no solution. □

Example 5. Solve the system

$$\begin{aligned} x - 2y + z &= 3 \\ 2x + 2y + 6z &= 14 \\ x + y + 3z &= 7. \end{aligned}$$

Solution. This is nearly the same as the previous example. After we eliminate x from the second and third equations, we have

$$\begin{aligned} x - \ 2y + \ \ \ z &= 3 \\ 6y + \ 4z &= 8 \\ 3y + \ 2z &= 4. \end{aligned}$$

Eliminating y from the third equation gives

$$\begin{aligned} x - \ 2y + \ \ \ z &= 3 \\ 6y + \ 4z &= 8 \\ 0y + \ 0z &= 0. \end{aligned}$$

This time the third equation has lots of solutions: y and z can be anything. The second equation is to be interpreted as giving a relation between y and z that must be satisfied if they are to be part of a solution. We'll take $y = -\frac{2}{3}z + \frac{4}{3}$ and let z be a free parameter. From the first equation, we'll then have that $x = 2y - z + 3 = -\frac{4}{3}z + \frac{8}{3} - z + 3$ or $x = -\frac{7}{3}z + \frac{17}{3}$. So, the solution set has the parametric equations

$$\begin{aligned} x &= -\tfrac{7}{3}t + \tfrac{17}{3} \\ y &= -\tfrac{2}{3}t + \tfrac{4}{3} \\ z &= t \end{aligned}$$

which you might recognize as the parametric equations of a line in three dimensional space. □

In this example, each of the original equations represented a plane in three dimensional space. The solution set was the set where these planes intersect and turned out to be a line. Any linear system involving only three variables can be thought of as representing the intersection of planes. Planes can intersect in only three ways: in a plane (this happens only if all of the planes are the same), in a line, or in a point. They can also fail to have any points in common.

Example **6.** Consider the two equations in four variables

$$\begin{aligned} 2x + 3y + z - 2w &= 5 \\ x - 2y + 4z + 3w &= 2. \end{aligned}$$

Solution. Again we'll eliminate x from one of the equations but to avoid dealing with fractions we'll eliminate x from the first equation by subtracting two times the second equation from it. This gives

$$\begin{aligned} 7y - 7z - 8w &= 1 \\ x - 2y + 4z + 3w &= 2. \end{aligned}$$

Further eliminations are not possible. The first equation gives a relationship between y, z, and w that must hold if they are to be part of a solution namely

$$y = z + \tfrac{8}{7}w + \tfrac{1}{7}.$$

Following the idea of the previous example, we let z and w be free parameters say $z = t$ and $w = s$. The second equation, $x = 2y - 4z - 3w + 2$, can then be written in terms of s and t:

$$x = 2[t + \tfrac{8}{7}s + \tfrac{1}{7}] - 4t - 3s + 2 = -2t - \tfrac{5}{7}s + \tfrac{16}{7}.$$

So, the solution is

$$\begin{aligned} x &= -2t - \tfrac{5}{7}s + \tfrac{16}{7} \\ y &= t + \tfrac{8}{7}s + \tfrac{1}{7} \\ z &= t \\ w &= s. \end{aligned}$$

It is understood that s and t are free parameters that can take on any real values. Since there are exactly two free parameters, we would say that the solution set is two dimensional and would geometrically think of it as being a plane. □

Example 7. Consider the overdetermined system

$$\begin{aligned} 3x + 4y &= 5 \\ x - 5y &= 4 \\ 2x + y &= -1. \end{aligned}$$

It is called overdetermined because there are more equations than unknowns.

Solution. We can eliminate x from the second equation by multiplying the first equation by $\frac{1}{3}$ and subtracting the result from the second equation. Similarly, if we multiply the first equation by $\frac{2}{3}$ and subtract the result from the third equation we will eliminate the x from the third equation. This gives

$$\begin{aligned} 3x + \qquad 4y &= 5 \\ (-5 - \tfrac{4}{3})y &= 4 - \tfrac{5}{3} \\ (1 - \tfrac{8}{3})y &= -1 - \tfrac{10}{3} \end{aligned}$$

or

$$\begin{aligned} 3x + 4y &= 5 \\ -\tfrac{19}{3}\, y &= \tfrac{7}{3} \\ -\tfrac{5}{3}\, y &= -\tfrac{13}{3}. \end{aligned}$$

The last two equations are clearly incompatible. Therefore, the system has no solution. This is not surprising because we are requiring two variables x and y to satisfy *three* conditions (equations). Unless the conditions are somehow related, we will not be able to find x and y. □

Example 8. Solve the system

$$\begin{aligned} 3x + 2y &= 10 \\ x - 4y &= -6 \\ 2x + y &= 6. \end{aligned}$$

Solution. Subtracting $\frac{1}{3}$ of the first equation from the second and $\frac{2}{3}$ of the first equation from the third, we eliminate x from these equations and end up with the system

$$\begin{aligned} 3x + \qquad 2y &= 10 \\ (-4 - \tfrac{2}{3})y &= -6 - \tfrac{10}{3} \\ (1 - \tfrac{4}{3})y &= 6 - \tfrac{20}{3} \end{aligned}$$

or

$$\begin{aligned} 3x + \qquad 2y &= 10 \\ -\tfrac{14}{3}y &= -\tfrac{28}{3} \\ -\tfrac{1}{3}y &= -\tfrac{2}{3}. \end{aligned}$$

Now the last two equations each give $y = 2$. From the first equation we conclude that $x = 2$. Therefore, the only solution of this system is $x = y = 2$. □

In the previous example it was not necessary to use Op. 3 – to interchange equations – to carry out the elimination. However, there are systems that require this.

Example **9.** Find the solution set of

$$\begin{aligned} 2x - y + z &= 1 \\ 2x - y + 2z &= 0 \\ 4x + 2y - z &= 2. \end{aligned}$$

Solution. If we subtract the first equation from the second and two times the first from the third, we get

$$\begin{aligned} 2x - y + z &= 1 \\ z &= -1 \\ 4y - 3z &= 0. \end{aligned}$$

This can be solved immediately or we can interchange the second and third equations to have

$$\begin{aligned} 2x - \quad y + \quad z &= 1 \\ 4y - 3z &= 0 \\ z &= -1. \end{aligned}$$

In either case we have the solution $z = -1$, $y = -\frac{3}{4}$, $x = \frac{5}{8}$. □

Remark. In the computer implementation of Gaussian Elimination one often performs interchanges to help insure accuracy. The rule of thumb is: If all coefficients are about the same order of magnitude, then the pivot equation for the unknown x_k should be the remaining equation whose coefficient of x_k has largest the absolute value. The term "remaining equations" refers to those equations from which $x_1, x_2, \ldots, x_{k-1}$ have been eliminated. □

1.2. EXERCISES

1. Use the first equation in each system as the pivot equation for x and eliminate x from the remaining equations

(a)
$$\begin{aligned} 3x + 2y &= 10 \\ 4x + y &= 2 \end{aligned}$$

(b)
$$\begin{aligned} x - y + z + w &= 4 \\ x + y + z &= 3 \\ 2x + z + 2w &= 0 \end{aligned}$$

(c)
$$\begin{aligned} x - y &= 4 \\ 2x + 3y &= 8 \\ 5x + 2y &= 20 \end{aligned}$$

(d)
$$\begin{aligned} 2x - y + 2z + w &= 0 \\ x + 3z &= 4 \\ 3x + y &= 1 \\ 2x + 4z + w &= 2 \end{aligned}$$

(e)
$$\begin{aligned} 2x - y &= 0 \\ -x + y + z &= 0 \\ 2x + z &= 5 \end{aligned}$$

(f)
$$\begin{aligned} 4x + 2y - z &= 1 \\ 2x + 2y + z &= 0 \\ -2x + 2z &= 0 \end{aligned}$$

2. Find all solutions of each of the following systems or show that there are no solutions.

(a)
$$\begin{aligned} x - y - 2z &= 0 \\ 2x + y - z &= 0 \\ x + 2y + z &= -1 \end{aligned}$$

(b)
$$\begin{aligned} 2x + 5y - 2z &= 5 \\ x + y &= 0 \\ 3x + 2z &= 4 \end{aligned}$$

(c)
$$\begin{aligned} x + y - z + 3w &= 6 \\ x + y + z + w &= 0 \\ x - y - z - w &= 0 \end{aligned}$$

(d)
$$\begin{aligned} 3x + 2y + z &= 0 \\ 2x - y + 3z &= 0 \\ x + 3y - 2z &= 0 \end{aligned}$$

(e)
$$\begin{aligned} x - y + z - w &= 0 \\ y + z &= 0 \\ x + 2y + z + w &= 0 \end{aligned}$$

(f)
$$\begin{aligned} 4x + 5y - 6z &= -1 \\ x + y + z &= 2 \\ 2x + 2z &= 2. \end{aligned}$$

3. Find the solution sets of the systems in Exercise 1.

4. A linear system in which all of the given constants on the right are equal to zero always has a solution. Explain.

5. Consider the system

$$\begin{aligned} 1.2969x + 0.8648y &= 0.8642 \\ 0.2161x + 0.1441y &= 0.1440. \end{aligned}$$

(a) A Martian observed that with $\hat{x} = 0.9911$ and $\hat{y} = -0.4870$ we have

$$\begin{aligned} 1.2969\hat{x} + 0.8648\hat{y} &= 0.86419999 \\ 0.2161\hat{x} + 0.1441\hat{y} &= 0.14400001. \end{aligned}$$

How close do you think $\hat{x}$ and $\hat{y}$ are to the true solution of the given system? Through how many decimal places do you think the true solution agrees with the given solution?

(b) Each of the equations in the original system represents a straight line in the x-y plane. Sketch these lines. Can you see where they cross?

(c) Use Gaussian Elimination to solve the given system. Do exact arithmetic; do not round-off any numbers.

1.3. A Closer Look at Gaussian Elimination

The Gaussian Elimination procedure for solving linear systems consists of two parts:

(i) the transformation of the given system into a triangular system;

(ii) the solution of the triangular system.

Part (i) is usually called *elimination* and part (ii) is called *backsolving.* Backsolving requires only elementary algebra while elimination entails the use of operations Op. 1, Op. 2, Op. 3.

Elimination accomplishes two things. It puts the left hand side of the system into the triangular form and it preprocesses the right hand side so it will be ready for backsolving. In some applications it is necessary to solve many systems all of which have the same left hand sides but whose right hand sides differ. This occurs for example in some modifications of Newton's method for non-linear systems (see Appendix B).

If we are faced with the problem of having to solve several systems all of which have the same left hand side, we can save a considerable amount of work by recording the steps used in carrying out elimination on the first system. We will use this information in subsequent systems to preprocess the right hand sides. Recall that the basic step in eliminating the unknown x_k from the ith equation is to multiply the pivot equation for x_k by a suitable number and then subtract the result from the ith equation. The multiplier used is of course

$$\frac{\textit{coefficient of } x_k \textit{ in } i^{th} \textit{ equation}}{\textit{coefficient of } x_k \textit{ in the pivot equation}}.$$

The method we propose is to write this multiplier in the place formerly occupied by x_k in the ith equation. As a rule we will not record zero multipliers. A person who knows the basic steps of Gaussian Elimination will be able to use this information to preprocess any right hand side.

Example 1. Solve the systems

$$\begin{aligned} 2x + 3y &= 4 \\ 6x + \quad y &= 5 \end{aligned} \qquad \begin{aligned} 2x + 3y &= 1 \\ 6x + \quad y &= 5 \end{aligned} \qquad \begin{aligned} 2x + 3y &= 0 \\ 6x + \quad y &= 1. \end{aligned}$$

Solution. We should multiply the first equation by 3 and subtract from the second. On the first system this gives

$$\begin{aligned} 2x + 3y &= 4 \\ 3 \quad - 8y &= -7 \end{aligned} \tag{1}$$

Notice that we have recorded the multiplier 3. Now we must backsolve

$$\begin{aligned} 2x \ + 3y &= 4 \\ - 8y &= -7 \end{aligned} \tag{2}$$

for $y = \frac{7}{8}$, $x = \frac{11}{16}$.

For the second system we will have to backsolve a system whose left hand side is exactly like (2)'s but whose right hand side comes from preprocessing the right hand side. From (1) we see that the right hand side of the first equation is multiplied by 3 and the result is subtracted from the right hand side of the second equation. This gives

$$\begin{aligned} 2x \ + 3y &= 1 \\ - 8y &= 5 - 3{\cdot}1 = 2 \end{aligned}$$

So, $y = -\frac{1}{4}$ and $x = \frac{7}{8}$.

Finally, for the third system we have

$$\begin{aligned} 2x + 3y &= 0 \\ - 8y &= 1 - 3{\cdot}0 = 1. \end{aligned}$$

So, $y = -\frac{1}{8}$ and $x = \frac{3}{16}$. □

Example **2.** Solve the systems (compare to Example 3 of Section 1.2).

$$\begin{aligned} x + y - 2z &= 1 \\ x + 3y - z &= 0 \\ x + 3y + z &= 0 \end{aligned} \qquad \begin{aligned} x + y - 2z &= 1 \\ x + 3y - z &= 2 \\ x + 3y + z &= 3 \end{aligned} \qquad \begin{aligned} x + y - 2z &= 4 \\ x + 3y - z &= 3 \\ x + 3y + z &= 5. \end{aligned}$$

Solution. We'll solve the first system. To eliminate x in the second equation we multiply the first equation by 1 and subtract the result from the second equation. We record the multiplier 1 by writing it in the place formerly occupied by the variable x in the second equation. Thus, we have

$$\begin{array}{llll} x + & y - & 2z = & 1 \\ 1 & 2y + & z = & -1 \\ x + & 3y + & z = & 0. \end{array}$$

Similarly, to eliminate x from the third equation we multiply the first equation by 1 and subtract the result from the third. We record the multiplier 1 in the position occupied by the quantity eliminated.

$$\begin{array}{llll} x + & y - & 2z = & 1 \\ 1 & 2y + & z = & -1 \\ 1 & 2y + & 3z = & -1. \end{array}$$

Now, to eliminate y in the third equation we multiply the second equation by 1 and subtract the result from our current third equation. Again the multiplier 1 is saved in the position formerly occupied by y in the third equation

$$\begin{array}{llll} x + & y - & 2z = & 1 \\ 1 & 2y + & z = & -1 \\ 1 & 1 & 2z = & 0. \end{array} \tag{3}$$

The system to backsolve is

$$\begin{aligned} x + \quad y - \; 2z &= 1 \\ 2y + \quad z &= -1 \\ 2z &= 0. \end{aligned}$$

The solution is $z = 0$, $y = -\frac{1}{2}$, $x = \frac{3}{2}$.

Now, for the second system we already know that the system we will have to backsolve will have the form

$$\begin{aligned} x + \quad y - \; 2z &= a \\ 2y + \quad z &= b \\ 2z &= c \end{aligned}$$

where a, b, and c result from applying the elimination (preprocessing) steps to the original right hand sides of 1, 2, and 3. From the system (3) we see the necessary steps:

$$\begin{matrix} 1 & & 1 & & 1 & & 1 \\ 2 & \overset{(1)}{\rightarrow} & 1 & \overset{(2)}{\rightarrow} & 1 & \overset{(3)}{\rightarrow} & 1 \\ 3 & & 3 & & 2 & & 1 \end{matrix}$$

Step 1 is to multiply the right hand side of the first equation by 1 and subtract the result from the second. Step 2 is to multiply the right hand side of the first equation by 1 and subtract the result from the third. Step 3 is to multiply the new right hand side of the second equation 1 and subtract the result from the third. Thus, we must solve

$$\begin{aligned} x + \quad y - \; 2z &= 1 \\ 2y + \quad z &= 1 \\ 2z &= 1. \end{aligned}$$

The solution is $z = \frac{1}{2}$, $y = \frac{1}{4}$, $x = \frac{7}{4}$.

For the third system we have the following processing of the right hand side

$$\begin{matrix} 4 & & 4 & & 4 & & 4 \\ 3 & \overset{(1)}{\to} & -1 & \overset{(2)}{\to} & -1 & \overset{(3)}{\to} & -1 \\ 5 & & 5 & & 1 & & 2. \end{matrix}$$

We solve

$$\begin{aligned} x + \quad y - 2z &= 4 \\ 2y + \quad z &= -1 \\ 2z &= 2 \end{aligned}$$

to get $z = 1$, $y = -1$, $x = 7$. □

Example 3. Solve the systems

$$\begin{aligned} 2x_1 - x_2 \qquad\qquad &= 1 \\ -x_1 + 2x_2 - x_3 \qquad &= 0 \\ x_1 - x_2 + 3x_3 - x_4 &= 0 \\ x_2 \qquad + 2x_4 &= 0 \end{aligned} \qquad\qquad \begin{aligned} 2x_1 - x_2 \qquad\qquad &= 0 \\ -x_1 + 2x_2 - x_3 \qquad &= 1 \\ x_1 - x_2 + 3x_3 - x_4 &= 0 \\ x_2 \qquad + 2x_4 &= 0. \end{aligned}$$

Solution. The $-x_1$ is eliminated in the second equation by multiplying the first by $-\frac{1}{2}$ and subtracting the first equation from the second. We write

$$\begin{aligned} 2x_1 - x_2 \qquad\qquad &= 1 \\ -\tfrac{1}{2} \qquad \tfrac{3}{2}x_2 - x_3 \qquad &= \tfrac{1}{2} \\ x_1 - x_2 + 3x_3 - x_4 &= 0 \\ x_2 + \qquad 2x_4 &= 0. \end{aligned}$$

Now x_1 in the third equation is eliminated by multiplying the first equation by $\frac{1}{2}$ and subtracting it from the third equation. This gives

$$
\begin{array}{rrrrl}
2x_1 & - \quad x_2 & & & = 1 \\
-\frac{1}{2} & \frac{3}{2}x_2 & - \quad x_3 & & = \frac{1}{2} \\
\frac{1}{2} & -\frac{1}{2}\ x_2 & + \ 3x_3 & - \quad x_4 & = -\frac{1}{2} \\
 & x_2 & + & 2x_4 & = 0.
\end{array}
$$

To eliminate x_2 from the third and fourth equations we use multipliers of $-\frac{1}{3}$ and $\frac{2}{3}$ respectively, this gives

$$
\begin{array}{rrrrl}
2x_1 & - \quad x_2 & & & = 1 \\
-\frac{1}{2} & \frac{3}{2}x_2 & - \quad x_3 & & = \frac{1}{2} \\
\frac{1}{2} & -\frac{1}{3} & \frac{8}{3}x_3 & - \quad x_4 & = -\frac{1}{3} \\
 & x_2 & & + \ 2x_4 & = 0
\end{array}
$$

and

$$
\begin{array}{rrrrl}
2x_1 & - \quad x_2 & & & = 1 \\
-\frac{1}{2} & \frac{3}{2}x_2 & - \quad x_3 & & = \frac{1}{2} \\
\frac{1}{2} & -\frac{1}{3} & \frac{8}{3}x_3 & - \quad x_4 & = -\frac{1}{3} \\
 & \frac{2}{3} & \frac{2}{3}x_3 & + \ 2x_4 & = -\frac{1}{3}
\end{array}
$$

Finally, to eliminate x_3 from the fourth equation we multiply the third equation by $\frac{1}{4}$ and subtract the result from the fourth equation:

$$
\begin{array}{rrrrl}
2x_1 & - \quad x_2 & & & = 1 \\
-\frac{1}{2} & \frac{3}{2}x_2 & - \quad x_3 & & = \frac{1}{2} \\
\frac{1}{2} & -\frac{1}{3} & \frac{8}{3}x_3 & - \quad x_4 & = -\frac{1}{3} \\
 & \frac{2}{3} & \frac{1}{4} & \frac{9}{4}x_4 & = -\frac{1}{4}
\end{array}
\qquad (4)
$$

So, we must backsolve

$$\begin{aligned} 2x_1 - \quad x_2 \qquad\qquad &= 1 \\ \tfrac{3}{2}x_2 - \quad x_3 \qquad &= \tfrac{1}{2} \\ \tfrac{8}{3}x_3 - \quad x_4 &= -\tfrac{1}{3} \\ \tfrac{9}{4}x_4 &= -\tfrac{1}{4}. \end{aligned}$$

The solution is $x_4 = -\frac{1}{9}$, $x_3 = -\frac{1}{6}$, $x_2 = \frac{2}{9}$, $x_1 = \frac{11}{18}$.

For the second system, we first preprocess the right hand side. There are five steps, one for each multiplier we saved.

$$\begin{matrix} 0 & & 0 & & 0 & & 0 & & 0 & & 0 \\ 1 & & 1 & & 1 & & 1 & & 1 & & 1 \\ 0 & \rightarrow & 0 & \rightarrow & 0 & \rightarrow & \frac{1}{3} & \rightarrow & \frac{1}{3} & \rightarrow & \frac{1}{3} \\ 0 & & 0 & & 0 & & 0 & & -\frac{2}{3} & & -\frac{3}{4}. \end{matrix} \tag{5}$$

Now, we backsolve

$$\begin{aligned} 2x_1 - \quad x_2 \qquad\qquad &= 0 \\ \tfrac{3}{2}x_2 - \quad x_3 \qquad &= 1 \\ \tfrac{8}{3}x_3 - \quad x_4 &= \tfrac{1}{3} \\ \tfrac{9}{4}x_4 &= -\tfrac{3}{4}. \end{aligned} \tag{6}$$

Thus, $x_4 = -\frac{1}{3}$, $x_3 = 0$, $x_2 = \frac{2}{3}$, $x_1 = \frac{1}{3}$. □

In each of the above examples the kth equation was the pivot equation for the kth variable. In cases where this is not so it is useful to record the pivot equation for each variable. We will do this by writing the number k in a circle next to the pivot equation for x_k. The multipliers will still be recorded in the place formerly occupied by the eliminated variable. The following simple example illustrates this point.

Example 4. Solve the systems

$$\text{(a)}\quad \begin{aligned} x_2 + 2x_3 &= 1 \\ 2x_2 + 3x_3 &= 0 \\ x_1 + x_2 + x_3 &= 0 \end{aligned} \qquad\qquad \text{(b)}\quad \begin{aligned} x_2 + 2x_3 &= 0 \\ 2x_2 + 3x_3 &= 1 \\ x_1 + x_2 + x_3 &= 0. \end{aligned}$$

Solution 1. First, the third equation is the pivot equation for x_1. If we pick the first equation to be the pivot for x_2 and eliminate x_2 from the remaining equation, we write

$$\begin{aligned} x_2 + 2x_3 &= 1 & \textcircled{2} \\ 2 - \quad x_3 &= -2 & \textcircled{3} \\ x_1 + x_2 + x_3 &= 0. & \textcircled{1} \end{aligned} \tag{7}$$

Backsolving now consists of proceeding from equation $\textcircled{3}$ to equation $\textcircled{1}$:

$$x_3 = 2,\ x_2 = -3,\ x_1 = 1.$$

For system (b) we preprocess the right hand side using (7):

$$\begin{matrix} 0 & & 0 \\ 1 & \rightarrow & 1 \\ 0 & & 0 \end{matrix}$$

The one on the right hand side of the second equation was not affected by the multiplier two since the right hand side of the first equation is zero. Now, we backsolve

$$\begin{aligned} x_2 + 2x_3 &= 0 \\ -\ x_3 &= 1 \\ x_1 + x_2 + x_3 &= 0 \end{aligned}$$

to get $x_3 = -1,\ x_2 = 2,\ x_1 = -1.$ □

Solution 2. If we had chosen the second equation to be the pivot for x_2, system (a) would reduce to

$$\begin{array}{rrrl} \frac{1}{2} & & \frac{1}{2}x_3 = 1 & \quad ③ \\ & 2x_2 + & 3x_3 = 0 & \quad ② \\ x_1 + & x_2 + & x_3 = 0 & \quad ① \end{array} \qquad (8)$$

which also gives $x_3 = 2,\ x_2 = -3,\ x_1 = 1$.

For system (b) we preprocess the right hand side using (8):

$$\begin{array}{ccc} 0 & & -\frac{1}{2} \\ 1 & \rightarrow & 1 \\ 0 & & 0 \end{array}$$

and backsolve

$$\begin{array}{rrrl} & & \frac{1}{2}x_3 = -\frac{1}{2} \\ & 2x_2 + & 3x_3 = 1 \\ x_1 + & x_2 + & x_3 = 0 \end{array}$$

to get $x_3 = -1,\ x_2 = 2,\ x_1 = -1$ as before. □

1.3. EXERCISES

1. Solve the systems below:

(a)
$$\begin{array}{rrrl} 2x - & y & & = 1 \\ -x + & 2y - & z & = 0 \\ - & y + & 2z & = 0 \end{array}$$

(b)
$$\begin{array}{rrrl} 2x - & y & & = 0 \\ -x + & 2y - & z & = 1 \\ -y + & & 2z & = 0 \end{array}$$

(c)
$$\begin{array}{rrrl} 2x - & y & & = 0 \\ -x + & 2y - & z & = 0 \\ - & y + & 2z & = 1 \end{array}$$

(c)
$$\begin{array}{rrrl} 2x - & y & & = 1 \\ -x + & 2y - & z & = 2 \\ -y + & & 2z & = 3 \end{array}$$

2. Use the computations in Example 3 to solve the systems

$$\begin{aligned} 2x_1 - x_2 \quad\quad &= 0 \\ -x_1 + 2x_2 - x_3 \quad &= 0 \\ x_1 - x_2 + 3x_3 - x_4 &= 1 \\ x_2 + 2x_4 &= 0 \end{aligned} \qquad\qquad \begin{aligned} 2x_1 - x_2 \quad\quad &= 0 \\ -x_1 + 2x_2 - x_3 \quad &= 0 \\ x_1 - x_2 + 3x_3 - x_4 &= 0 \\ x_2 + 2x_4 &= 1. \end{aligned}$$

3. Solve the systems

(a)
$$\begin{aligned} 4x - y \quad\quad - w &= 1 \\ -x + 4y - z \quad\quad &= 0 \\ -y + 4z - w &= 0 \\ -x \quad - z + 4w &= 0 \end{aligned}$$

(b)
$$\begin{aligned} 4x - y \quad\quad - w &= 0 \\ -x + 4y - z \quad\quad &= 1 \\ -y + 4z - w &= 0 \\ -x \quad - z + 4w &= 0 \end{aligned}$$

(c)
$$\begin{aligned} 4x - y \quad\quad - w &= 0 \\ -x + 4y - z \quad\quad &= 0 \\ -y + 4z - w &= 1 \\ -x \quad - z + 4w &= 0 \end{aligned}$$

(d)
$$\begin{aligned} 4x - y \quad\quad - w &= 0 \\ -x + 4y - z \quad\quad &= 0 \\ -y + 4z - w &= 0 \\ -x \quad - z + 4w &= 1 \end{aligned}$$

(e)
$$\begin{aligned} 4x - y \quad\quad - w &= 1 \\ -x + 4y - z \quad\quad &= 1 \\ -y + 4z - w &= 0 \\ -x \quad - z + 4w &= 2 \end{aligned}$$

4. Solve the systems

(a)
$$\begin{aligned} 2x - y + z &= 1 \\ 2x - y + 2z &= 0 \\ 4x + 2y - z &= 0 \end{aligned}$$

(b)
$$\begin{aligned} 2x - y + z &= 0 \\ 2x - y + 2z &= 1 \\ 4x + 2y - z &= 0 \end{aligned}$$

(c)
$$\begin{aligned} 2x - y + z &= 0 \\ 2x - y + 2z &= 0 \\ 4x + 2y - z &= 1 \end{aligned}$$

(d)
$$\begin{aligned} 2x - y + z &= 3 \\ 2x - y + 2z &= -1 \\ 4x + 2y - z &= 2 \end{aligned}$$

CHAPTER 2 Matrices

In this chapter we introduce matrix notation for linear systems and given an overview of matrix algebra. We also briefly discuss inverses and determinants.

2.1. Matrices, Column Vectors, and Row Vectors

The names of the variables play no essential role in Gaussian Elimination. The coefficients of the variables and the numbers on the right hand sides of the equations are what determine the system and are what we manipulate to solve the system. It is common practice to isolate the roles of the coefficients, the variables, and the right hand sides by grouping them in an orderly and natural way. The coefficients of the variables are arranged in a rectangular array called a matrix. The variables and the right hand sides are arranged in "column vectors."

Example 1. The system

$$\begin{aligned} 2x - 3y + 4z &= 10 \\ 5x - y + z &= 2 \\ x - y + 2z &= 1 \end{aligned}$$

has the matrix-column vector representation

$$\begin{bmatrix} 2 & -3 & 4 \\ 5 & -1 & 1 \\ 1 & -1 & 2 \end{bmatrix} \begin{bmatrix} x \\ y \\ z \end{bmatrix} = \begin{bmatrix} 10 \\ 2 \\ 1 \end{bmatrix}. \qquad \square$$

Example 2. The system

$$\begin{aligned} x - y + 3z + w &= 5 \\ 2x + 3y + w &= 6 \\ 7x + z + w &= 0 \end{aligned}$$

has the representation

$$\begin{bmatrix} 1 & -1 & 3 & 1 \\ 2 & 3 & 0 & 1 \\ 7 & 0 & 1 & 1 \end{bmatrix} \begin{bmatrix} x \\ y \\ z \\ w \end{bmatrix} = \begin{bmatrix} 5 \\ 6 \\ 0 \end{bmatrix}. \qquad \square$$

Here are the formal definitions.

Definition. For integers n and k an *n by k matrix* (also written as "$n{\times}k$ matrix") is a rectangular array of numbers consisting of n horizontal rows and k vertical columns. The numbers that make-up the matrix are called its *entries*.

Definitions. A *column vector* of length n is an $n{\times}1$ matrix.

A *row vector* of length n is a $1{\times}n$ matrix.

Example **3.** In the 2×2 matrix $\begin{bmatrix} 1 & 2 \\ 3 & 4 \end{bmatrix}$ we have the column vectors $\begin{bmatrix} 1 \\ 3 \end{bmatrix}$ and $\begin{bmatrix} 2 \\ 4 \end{bmatrix}$ and the row vectors $[1 \quad 2]$ and $[3 \quad 4]$.

Example **4.** In the 3×4 matrix in Example 2, we have the four column vectors

$$\begin{bmatrix} 1 \\ 2 \\ 7 \end{bmatrix}, \begin{bmatrix} -1 \\ 3 \\ 0 \end{bmatrix}, \begin{bmatrix} 3 \\ 0 \\ 1 \end{bmatrix}, \text{ and } \begin{bmatrix} 1 \\ 1 \\ 1 \end{bmatrix}$$

and the three rows vectors $[1 \quad -1 \quad 3 \quad 1]$, $[2 \quad 3 \quad 0 \quad 1]$, and $[7 \quad 0 \quad 1 \quad 1]$. $\square$

Remark. As these examples show, we can view an $n{\times}k$ matrix as being either an ordered collection of k column vectors each of length n or an ordered collection of n row vectors each of length k.

Matrices will be denoted by capital letters. If A is an $n{\times}k$ matrix, then $A(i,j)$ or A_{ij} denotes the entry of A lying in the ith row from the top and j^{th} column from the left. For example $A(1,1)$ is the entry in the upper left corner and $A(n,k)$ is the entry in the lower right corner. $A(i,j)$ is called the *i-j^{th} entry of A.*

Example **5.** For the 3×2 matrix $B = \begin{bmatrix} 1 & 0 \\ 7 & -1 \\ 8 & 5 \end{bmatrix}$, we have $B(1,2) = 0$, $B(2,1) = 7$, $B(2,2) = -1$, and $B_{3,1} = 8$. The quantity $B(2,3)$ is undefined since B has only two columns.

Column vectors will be denoted by bold face lower case letters. If $\mathbf{x}$ is a column vector of length n, then x_i is the ith entry or i^{th} *coordinate* of $\mathbf{x}$ (from the left).

Example 6. Let $\mathbf{x} = \begin{bmatrix} 0 \\ 8 \\ 2 \end{bmatrix}$, $\mathbf{y} = \begin{bmatrix} -7 \\ 3 \end{bmatrix}$, and $\mathbf{z} = \begin{bmatrix} 1 \\ 3 \\ -1 \\ 0 \end{bmatrix}$, then $x_1 = 0$, $x_2 = 8$, $x_3 = 2$, $y_2 = 3$, $z_1 = 1$, $z_4 = 0$. The quantities x_0, y_3, and z_5 are not defined. □

Row vectors will be denoted by bold face lower case letters with a superscript t, e.g., $\mathbf{x}^t$, $\mathbf{b}^t$. One thinks of $\mathbf{x}$ as being a column vector and $\mathbf{x}^t$ as being its row vector counterpart. $\mathbf{x}^t$ is called the *transpose of* $\mathbf{x}$.

Example 7. Let $\mathbf{x}^t = [0 \quad 1]$, $\mathbf{y}^t = [1 \quad 2 \quad 3]$, and $\mathbf{z}^t = [1 \quad 0 \quad 1 \quad 0]$. Then, $x_1^t = 0$, $x_2^t = 1$, $y_2^t = 2$, $z_4^t = 0$, x_3^t, y_5^t, z_5^t are not defined.

The set of all ordered n-tuples of real numbers will be denoted by $\mathbb{R}^n$. That is,

$$\mathbb{R}^n = \{(x_1,x_2,...,x_n)\colon \textit{ each } x_i \textit{ is a real number}\}.$$

The set of real numbers will be denoted by $\mathbb{R}$.

We will often identify column vectors with members of $\mathbb{R}^n$ in the natural way: $\mathbf{x} = (x_1,...,x_n)$. For example (1,2) and $\begin{bmatrix} 1 \\ 2 \end{bmatrix}$ are both thought of as the point in the Cartesian plane whose first coordinate is 1 and whose second is 2.

Definition. Two matrices are *equal* if and only if they are equal entry wise. This notion of equality refers to row vectors and column vectors also.

The representation of a system of linear equations in terms of a matrix of coefficients, a column vector of unknowns, and a column vector of known numbers tells us how the product of a matrix and a column vector should be defined. For if

$$\begin{aligned} ax + by &= \alpha \\ cx + dy &= \beta \end{aligned}$$

is to mean the same as

$$\begin{bmatrix} a & b \\ c & d \end{bmatrix}\begin{bmatrix} x \\ y \end{bmatrix} = \begin{bmatrix} \alpha \\ \beta \end{bmatrix},$$

then $\begin{bmatrix} a & b \\ c & d \end{bmatrix}\begin{bmatrix} x \\ y \end{bmatrix}$ *must* be the column vector $\begin{bmatrix} ax+by \\ cx+dy \end{bmatrix}$. In this case, we have

$$\begin{bmatrix} ax+by \\ cx+dy \end{bmatrix} = \begin{bmatrix} \alpha \\ \beta \end{bmatrix}$$

which means $ax + by = \alpha$ and $cx + dy = \beta$.

Matrix-Vector Products

We will define matrix-vector products in terms of the following important notion.

Definition. Let $\mathbf{x}$ and $\mathbf{y}$ be in $\mathbb{R}^n$. The *inner product* of $\mathbf{x}$ and $\mathbf{y}$ is defined by

$$\mathbf{x}^t\mathbf{y} = x_1y_1 + x_2y_2 + \cdots + x_ny_n.$$

This is also called the *dot product* or the *scalar product* and is sometimes denoted by $\mathbf{x}\cdot\mathbf{y}$.

Observe that $\mathbf{x}^t\mathbf{y} = \mathbf{y}^t\mathbf{x}$ holds. For example, $[1 \quad 2]\begin{bmatrix} 3 \\ 4 \end{bmatrix} = 3 + 8 = [3 \quad 4]\begin{bmatrix} 1 \\ 2 \end{bmatrix}$.

Here is the formal definition of the matrix-column vector product.

Definition. Let A be an $n\times k$ matrix whose row vectors are $\mathbf{r}_1^t,\mathbf{r}_2^t,\ldots,\mathbf{r}_n^t$ and let $\mathbf{y}$ be a column vector of length k, then the product $A\mathbf{y}$ is the column vector of length n whose i^{th} coordinate is $\mathbf{r}_i^t\mathbf{y}$ for $i = 1,2,\ldots,n$.

In short

$$A\mathbf{y} = \begin{bmatrix} \mathbf{r}_1^t\mathbf{y} \\ \mathbf{r}_2^t\mathbf{y} \\ \vdots \\ \mathbf{r}_n^t\mathbf{y} \end{bmatrix}.$$

Example 8. Let $\mathbf{x} = \begin{bmatrix} 1 \\ 3 \\ 5 \end{bmatrix}$, $\mathbf{y} = \begin{bmatrix} 0 \\ 2 \\ 1 \end{bmatrix}$, and $\mathbf{z} = \begin{bmatrix} -1 \\ 7 \\ 6 \end{bmatrix}$. Find $\mathbf{x}^t\mathbf{y}$, $\mathbf{y}^t\mathbf{z}$, $\mathbf{z}^t\mathbf{x}$.

Solution.

$$\mathbf{x}^t\mathbf{y} = [1 \quad 3 \quad 5]\begin{bmatrix} 0 \\ 2 \\ 1 \end{bmatrix} = 1\cdot 0 + 3\cdot 2 + 5\cdot 1 = 11.$$

$$\mathbf{y}^t\mathbf{z} = 0\cdot(-1) + 2\cdot 7 + 1\cdot 6 = 20.$$

$$\mathbf{z}^t\mathbf{x} = (-1)\cdot 1 + 7\cdot 3 + 6\cdot 5 = 50.$$

□

Example 9. Let $\mathbf{x}$, $\mathbf{y}$, and $\mathbf{z}$ be as in Example 8 and let

$$A = \begin{bmatrix} 1 & 2 & 0 \\ 8 & 0 & 1 \\ 6 & 5 & 2 \end{bmatrix} \text{ and } B = \begin{bmatrix} 1 & 7 & -1 \\ 0 & 2 & 4 \end{bmatrix}$$

find $A\mathbf{x}$, $A\mathbf{y}$, $B\mathbf{y}$, and $B\mathbf{z}$.

Solution. For $A\mathbf{x}$ we must compute

$$[1 \quad 2 \quad 0]\begin{bmatrix} 1 \\ 3 \\ 5 \end{bmatrix}, [8 \quad 0 \quad 1]\begin{bmatrix} 1 \\ 3 \\ 5 \end{bmatrix}, \text{ and } [6 \quad 5 \quad 2]\begin{bmatrix} 1 \\ 3 \\ 5 \end{bmatrix}.$$

The results are 7, 13, and 31; so

$$A\mathbf{x} = \begin{bmatrix} 7 \\ 13 \\ 31 \end{bmatrix}.$$

For $A\mathbf{y}$ we have

$$\begin{bmatrix} 1 & 2 & 0 \\ 8 & 0 & 1 \\ 6 & 5 & 2 \end{bmatrix}\begin{bmatrix} 0 \\ 2 \\ 1 \end{bmatrix} = \begin{bmatrix} 0\cdot 1+2\cdot 2+0\cdot 1 \\ 8\cdot 0+0\cdot 2+1\cdot 1 \\ 6\cdot 0+5\cdot 2+2\cdot 1 \end{bmatrix} = \begin{bmatrix} 4 \\ 1 \\ 12 \end{bmatrix}.$$

For $B\mathbf{y}$ we expect a column vector of length 2 as the result. We have

$$\begin{bmatrix} 1 & 7 & -1 \\ 0 & 2 & 4 \end{bmatrix}\begin{bmatrix} 0 \\ 2 \\ 1 \end{bmatrix} = \begin{bmatrix} 1\cdot 0+7\cdot 2+(-1)\cdot 1 \\ 0\cdot 0+2\cdot 2+4\cdot 1 \end{bmatrix} = \begin{bmatrix} 13 \\ 8 \end{bmatrix}.$$

For Bz we have

$$\begin{bmatrix} 1 & 7 & -1 \\ 0 & 2 & 4 \end{bmatrix}\begin{bmatrix} -1 \\ 7 \\ 6 \end{bmatrix} = \begin{bmatrix} 1\cdot(-1)+7\cdot7+(-1)\cdot6 \\ 0\cdot(-1)+2\cdot7+4\cdot6 \end{bmatrix} = \begin{bmatrix} 42 \\ 38 \end{bmatrix}.$$ □

Remark. One can view an $n\times k$ matrix as a machine that turns column vectors of length k into column vectors of length n. Note that the length of the input vectors is equal to the number of columns in the matrix and that the length of the output vectors is equal to the number of rows in the matrix. □

A matrix is uniquely determined by its action on some very special input vectors.

Definitions. 1. The *standard unit basis vectors in* $\mathbb{R}^2$ are the vectors $\mathbf{e}_1 = \begin{bmatrix} 1 \\ 0 \end{bmatrix}$ and $\mathbf{e}_2 = \begin{bmatrix} 0 \\ 1 \end{bmatrix}$.

2. The *standard unit basis vectors in* $\mathbb{R}^3$ are the vectors $\mathbf{e}_1 = \begin{bmatrix} 1 \\ 0 \\ 0 \end{bmatrix}$, $\mathbf{e}_2 = \begin{bmatrix} 0 \\ 1 \\ 0 \end{bmatrix}$, and $\mathbf{e}_3 = \begin{bmatrix} 0 \\ 0 \\ 1 \end{bmatrix}$.

3. In $\mathbb{R}^n$ the *standard unit basis vectors* are $\mathbf{e}_1,\ldots,\mathbf{e}_n$ where $\mathbf{e}_i$ is the vector with a one in the ith coordinate and zeros everywhere else.

Example 10. Let $A = \begin{bmatrix} a & d & g \\ b & e & h \\ c & f & i \end{bmatrix}$. Verify that $A\mathbf{e}_1$, $A\mathbf{e}_2$ and $A\mathbf{e}_3$ are the columns of A.

Solution.

$$A = \begin{bmatrix} a & d & g \\ b & e & h \\ c & f & i \end{bmatrix}\begin{bmatrix} 1 \\ 0 \\ 0 \end{bmatrix} = \begin{bmatrix} a\cdot1+d\cdot0+g\cdot0 \\ b\cdot1+e\cdot0+h\cdot0 \\ c\cdot1+f\cdot0+i\cdot0 \end{bmatrix} = \begin{bmatrix} a \\ b \\ c \end{bmatrix}.$$

The verifications of $A\mathbf{e}_2$ and $A\mathbf{e}_3$ are similar. □

In general we have

Theorem 1. If A is an $n\times k$ matrix, then $A = [A\mathbf{e}_1 \quad A\mathbf{e}_2 \quad \cdots \quad A\mathbf{e}_k]$.

In later chapters we will deal more completely with matrices as functions. For now, however, we will simply use the dimensions of the rows and columns of matrices as a check on our work.

Example 11. Let $A = \begin{bmatrix} 0 & 0 & 0 & 0 \\ 1 & 2 & 3 & 0 \end{bmatrix}$ and $B = \begin{bmatrix} 0 & 0 \\ 1 & 2 \\ 1 & 4 \end{bmatrix}$ and let

$$\mathbf{x} = \begin{bmatrix} 8 \\ 1 \end{bmatrix}, \mathbf{y} = \begin{bmatrix} 4 \\ 1 \\ 3 \end{bmatrix}, \text{ and } \mathbf{z} = \begin{bmatrix} 1 \\ 0 \\ 4 \\ 3 \end{bmatrix}.$$

Which of the products $A\mathbf{x}$, $A\mathbf{y}$, $A\mathbf{z}$, $B\mathbf{x}$, $B\mathbf{y}$, $B\mathbf{z}$ are defined? Of those that are defined what is the length of the resulting column vector?

Solution. A is 2×4 so $A\mathbf{w}$ is defined only if $\mathbf{w}$ is a column vector of length 4. The result is a column vector of length 2. Therefore, $A\mathbf{z}$ is defined. B is 3×2 so for $B\mathbf{w}$ to be defined $\mathbf{w}$ must be of length 2 and the result will be of length 3. $B\mathbf{x}$ is the only other product defined. □

We will occasionally have to deal with products of the form

$$\mathbf{x}^t A$$

where A is an $n \times k$ matrix and $\mathbf{x}^t$ is a row vector of length n.

Definition. Let A be an $n \times k$ matrix whose column vectors are $\mathbf{c}_1, \mathbf{c}_2, \ldots, \mathbf{c}_k$ and let $\mathbf{x}^t$ be a row vector of length n, then $\mathbf{x}^t A$ is the row vector of length k whose entries are $\mathbf{x}^t\mathbf{c}_1, \ldots, \mathbf{x}^t\mathbf{c}_k$:

$$\mathbf{x}^t A = [\mathbf{x}^t\mathbf{c}_1 \quad \mathbf{x}^t\mathbf{c}_2 \quad \cdots \quad \mathbf{x}^t\mathbf{c}_k].$$

Definition. Let A be an $n \times k$ matrix whose column vectors are $\mathbf{c}_1, \mathbf{c}_2, \ldots, \mathbf{c}_k$, then the *transpose* of A is the $k \times n$ matrix whose row vectors are $\mathbf{c}_1^t, \ldots, \mathbf{c}_k^t$:

$$A = [\mathbf{c}_1 \quad \mathbf{c}_2 \quad \cdots \quad \mathbf{c}_k] \text{ and } A^t = \begin{bmatrix} \mathbf{c}_1^t \\ \mathbf{c}_2^t \\ \vdots \\ \mathbf{c}_k^t \end{bmatrix}.$$

Example 12. Let $B = \begin{bmatrix} 1 & 3 \\ 2 & 0 \\ 4 & 5 \end{bmatrix}$ and $\mathbf{x} = \begin{bmatrix} x \\ y \end{bmatrix}$. Compute $B\mathbf{x}$, $\mathbf{x}^t B^t$ and verify that $(B\mathbf{x})^t = \mathbf{x}^t B^t$.

Solution. $B\mathbf{x} = \begin{bmatrix} x+3y \\ 2x \\ 4x+5y \end{bmatrix}$, $\mathbf{x}^t B^t = [x \quad y]\begin{bmatrix} 1 & 2 & 4 \\ 3 & 0 & 5 \end{bmatrix} = [x+3y \quad 2x \quad 4x+5y]$. By inspection $\mathbf{x}^t B^t = (B\mathbf{x})^t$. □

The computation of Example 12 is true in general.

Theorem 2. If A is an $n \times k$ matrix and $\mathbf{x}$ is in $\mathbb{R}^k$, then

$$(A\mathbf{x})^t = \mathbf{x}^t A^t.$$

Properties of the Inner Product

The concept of inner product of vectors in $\mathbb{R}^2$ or $\mathbb{R}^3$ is generally introduced in Calculus courses. We recall here some of the basic properties.

(1) The *magnitude* or *length* of a vector is $|\mathbf{x}| = \sqrt{\mathbf{x}^t\mathbf{x}}$. (This is not to be confused with the other notion of length introduced earlier which is simply the number of coordinates.) A vector with $|\mathbf{x}| = 1$ is called a *unit* vector.

(2) The *distance* between $\mathbf{x}$ and $\mathbf{y}$ is $|\mathbf{x} - \mathbf{y}|$ where the difference is defined coordinate wise.

(3) The *angle* between $\mathbf{x}$ and $\mathbf{y}$ is implicitly defined by

$$\mathbf{x}^t\mathbf{y} = |\mathbf{x}||\mathbf{y}|\cos\theta, \text{ where } 0 \le \theta \le \pi.$$

(4) $\mathbf{x}$ and $\mathbf{y}$ are *orthogonal* or *perpendicular* if $\mathbf{x}^t\mathbf{y} = 0$.

(5) The *projection* of $\mathbf{x}$ onto $\mathbf{y}$ is the vector

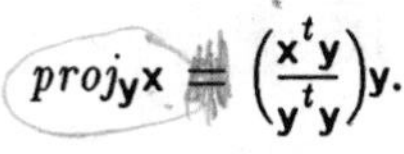

$$proj_{\mathbf{y}}\mathbf{x} = \left(\frac{\mathbf{x}^t\mathbf{y}}{\mathbf{y}^t\mathbf{y}}\right)\mathbf{y}.$$

The notion of perpendicularity (4) will arise quite often.

Example 13. Find the equation of the plane passing through $\mathbf{p}$ and perpendicular to $\mathbf{n}$.

Solution. A point $\mathbf{x}$ is on the desired plane only if the straight line motion from $\mathbf{p}$ to $\mathbf{x}$ is perpendicular to n. This geometric statement has the algebraic translation as

$$\mathbf{n}^t(\mathbf{x}-\mathbf{p}) = 0.$$

So, if we had $\mathbf{p} = (1,2,3)$ and $\mathbf{n} = (4,-5,6)$, then our equation would read

$$[4 \quad -5 \quad 6]\begin{bmatrix} x-1 \\ y-2 \\ z-3 \end{bmatrix} = 0$$

or

$$4(x-1) - 5(y-2) + 6(z-3) = 0. \qquad \square$$

Example 14. Let $\mathbf{x} = \begin{bmatrix} 1 \\ 2 \\ 1 \end{bmatrix}$, $\mathbf{y} = \begin{bmatrix} 1 \\ -1 \\ 1 \end{bmatrix}$, $\mathbf{z} = \begin{bmatrix} 0 \\ 2 \\ 1 \end{bmatrix}$. Find the angles between all pairs of vectors.

Solution. We have $|\mathbf{x}| = \sqrt{1+4+1} = \sqrt{6}$, $|\mathbf{y}| = \sqrt{1+1+1} = \sqrt{3}$, $|\mathbf{z}| = \sqrt{5}$ and $\mathbf{x}^t\mathbf{y} = 1 - 2 + 1 = 0$. So the angle between x and y satisfies

$$0 = \mathbf{x}^t\mathbf{y} = \cos\theta|\mathbf{x}||\mathbf{y}| \text{ or } \theta = \frac{\pi}{2}.$$

Since $\mathbf{x}^t\mathbf{z} = 4+1 = 5$, we have that the angle between x and z satisfies

$$5 = \mathbf{x}^t\mathbf{z} = \sqrt{6}\sqrt{5}\cos\theta \text{ so } \theta = \cos^{-1}\left(\frac{5}{\sqrt{30}}\right).$$

Finally, for the angle between y and z

$$-1 = \mathbf{y}^t\mathbf{z} = \sqrt{15}\cos\theta \text{ so } \theta = \cos^{-1}\left(\frac{-1}{\sqrt{15}}\right).$$

In summary, x and y are perpendicular, x and z form an acute angle since the cosine is positive, and y and z form an obtuse angle. $\square$

2.1. EXERCISES

1. Write each system in Exercise Set 1.2 in matrix-column vector form.

2. Let $\mathbf{x} = \begin{bmatrix} 1 \\ 2 \\ 3 \\ 1 \end{bmatrix}$, $\mathbf{y} = \begin{bmatrix} 0 \\ 1 \\ 7 \\ 5 \end{bmatrix}$, $\mathbf{z} = \begin{bmatrix} -1 \\ 4 \\ -1 \\ 0 \end{bmatrix}$. Find $\mathbf{x}^t\mathbf{y}$, $\mathbf{x}^t\mathbf{z}$, $\mathbf{z}^t\mathbf{y}$.

3. Let $A = \begin{bmatrix} 4 & 3 \\ 0 & 1 \\ 1 & 0 \end{bmatrix}$, $B = \begin{bmatrix} -1 & 6 \\ 7 & -2 \\ 0 & 0 \end{bmatrix}$, and $C = \begin{bmatrix} 0 & 4 \\ 0 & 8 \\ 0 & 7 \end{bmatrix}$.

   Let $\mathbf{w} = \begin{bmatrix} 4 \\ 1 \end{bmatrix}$. (a) Find $A\mathbf{w}$, $B\mathbf{w}$, $C\mathbf{w}$. (b) Find $(A\mathbf{w})^t(B\mathbf{w})$, $(A\mathbf{w})^t(C\mathbf{w})$, $(B\mathbf{w})^t(C\mathbf{w})$.

4. Let M be the matrix whose row vectors (from top to bottom) are $\mathbf{x}^t$, $\mathbf{y}^t$, $\mathbf{z}^t$ where $\mathbf{x}$, $\mathbf{y}$, and $\mathbf{z}$ are given in Problem 2. Find $M\mathbf{x}$, $M\mathbf{y}$, $M\mathbf{z}$.

5. For matrix A of Problem 3 the system $\mathbf{x}^tA = \mathbf{0}^t$ represents the intersection of two planes. What is the angle between the normals of the planes?

6. Let A be as in Problem 3. Compute $\mathbf{e}_1^tA$, $\mathbf{e}_2^tA$, and $\mathbf{e}_3^tA$. State a version of Theorem 1 that relates the rows of a matrix to these special products.

7. Let $A = \begin{bmatrix} 2 & 1 \\ 1 & 2 \end{bmatrix}$ and $\mathbf{x} = \begin{bmatrix} x \\ y \end{bmatrix}$. (a) Find the products $\mathbf{x}^t(A\mathbf{x})$ and $(\mathbf{x}^tA)\mathbf{x}$ and show that they are equal. {Compute the product in the parentheses first.} (b) Can you find values of x and y for which the product found in part (a) is negative?

8. Let $B = \begin{bmatrix} 1 & 1 \\ 1 & 1 \end{bmatrix}$ can you find $\mathbf{x}$ so that $\mathbf{x}^tB\mathbf{x} < 0$?

9. If A and B are such that the product AB is defined, then $(AB)^t = B^tA^t$. Verify this in the case that A and B are 2×2 matrices.

2.2. Matrix Algebra

We define addition and multiplication of matrices and state the important algebraic rules governing these operations.

There are a few basic ways of creating new vectors and matrices out of old ones. The simplest way is *scaling.*

Definition. Let A be a matrix and s a real number, then sA denotes the matrix obtained by multiplying each entry of A by s. This definition holds for row and column vectors since they are special matrices.

Example 1. Let $A = \begin{bmatrix} 1 & 3 & 8 \\ 2 & 5 & 9 \end{bmatrix}$, $\mathbf{x} = \begin{bmatrix} 4 \\ 5 \end{bmatrix}$, $\mathbf{y} = \begin{bmatrix} 2 \\ 3 \end{bmatrix}$. Find $2A$, $3\mathbf{x}$, and $-2\mathbf{y}^t$.

Solution. $$2A = \begin{bmatrix} 2 & 6 & 16 \\ 4 & 10 & 18 \end{bmatrix},\ 3\mathbf{x} = \begin{bmatrix} 12 \\ 15 \end{bmatrix},\ -2\mathbf{y}^t = [-4 \quad -6].$$ □

Often, a common factor is taken out of a matrix or vector. For example

Definition. Let A and B be $n{\times}k$ matrices. The *sum* $A+B$ is the $n{\times}k$ matrix whose entries are the sums of the entries of A and B.

Example 2. Let $A = \begin{bmatrix} 1 & 5 \\ 4 & 3 \end{bmatrix}$, $B = \begin{bmatrix} 2 & 1 \\ 3 & 0 \end{bmatrix}$, $\mathbf{x} = \begin{bmatrix} 4 \\ 5 \end{bmatrix}$ and $\mathbf{y} = \begin{bmatrix} -1 \\ -2 \end{bmatrix}$, then

$$A+B = \begin{bmatrix} 1+2 & 5+1 \\ 4+3 & 3+0 \end{bmatrix} = \begin{bmatrix} 3 & 6 \\ 7 & 3 \end{bmatrix} \text{ and } \mathbf{x} + \mathbf{y} = \begin{bmatrix} 4-1 \\ 5-2 \end{bmatrix} = \begin{bmatrix} 3 \\ 3 \end{bmatrix}$$ □

Since scaling and addition are defined coordinate wise in terms of the standard operations of arithmetic, all of the familiar laws of arithmetic hold for these operations. The matrix of all zeros is called the *zero matrix* and is denoted by O. The column vector of all zeros is denoted by $\mathbf{0}$.

Now, how is the product of matrices defined? We will build upon the definition of the matrix-column vector product.

Definition. Let A be an $n{\times}k$ matrix and let B be the $k{\times}p$ matrix whose column vectors are $\mathbf{b}_1,\ldots,\mathbf{b}_p$. Then, the *product* AB is the $n{\times}p$ matrix whose column vectors are $A\mathbf{b}_1$, $A\mathbf{b}_2,\ldots,A\mathbf{b}_p$. That is,

$$AB = A[\mathbf{b}_1 \quad \mathbf{b}_2 \quad \cdots \quad \mathbf{b}_p] = [A\mathbf{b}_1 \quad A\mathbf{b}_2 \quad \cdots \quad A\mathbf{b}_p].$$

This product can be viewed as the simultaneous multiplication of the columns of B by A. For the product AB to be defined the number of rows of B must equal the number of columns of A.

***Example* 3.** Let $A = \begin{bmatrix} 2 & 4 & 5 \\ 1 & 3 & 6 \end{bmatrix}$ and $B = \begin{bmatrix} 7 & 0 \\ 8 & -1 \\ 9 & -2 \end{bmatrix}$. Find AB and BA.

Solution.

$$AB = \begin{bmatrix} 2 & 4 & 5 \\ 1 & 3 & 6 \end{bmatrix}\begin{bmatrix} 7 & 0 \\ 8 & -1 \\ 9 & -2 \end{bmatrix} = \left[A\begin{bmatrix} 7 \\ 8 \\ 9 \end{bmatrix} A\begin{bmatrix} 0 \\ -1 \\ -2 \end{bmatrix}\right]$$

$$= \begin{bmatrix} 14+32+45 & 0-4-10 \\ 7+24+54 & 0-3-12 \end{bmatrix} = \begin{bmatrix} 91 & -14 \\ 85 & -15 \end{bmatrix}.$$

$$BA = \begin{bmatrix} 7 & 0 \\ 8 & -1 \\ 9 & -2 \end{bmatrix}\begin{bmatrix} 2 & 4 & 5 \\ 1 & 3 & 6 \end{bmatrix} = \left[B\begin{bmatrix} 2 \\ 1 \end{bmatrix} B\begin{bmatrix} 4 \\ 3 \end{bmatrix} B\begin{bmatrix} 5 \\ 6 \end{bmatrix}\right]$$

$$= \begin{bmatrix} 14 & 28 & 35 \\ 15 & 29 & 34 \\ 16 & 30 & 33 \end{bmatrix}$$

□

Notice that BA and AB are not equal, they do not even have the same dimensions!

***Example* 4.** Let $\mathbf{x} = \begin{bmatrix} 3 \\ 5 \end{bmatrix}$ and $\mathbf{y}^t = [1 \quad 6]$ and view them as 2×1 and 1×2 matrices respectively. Find $\mathbf{x}\mathbf{y}^t$ and $\mathbf{y}^t\mathbf{x}$.

Solution.

$$\mathbf{x}\mathbf{y}^t = \begin{bmatrix} 3 \\ 5 \end{bmatrix}[1 \quad 6] = \left[\begin{bmatrix} 3 \\ 5 \end{bmatrix}1 \begin{bmatrix} 3 \\ 5 \end{bmatrix}6\right] = \begin{bmatrix} 3 & 18 \\ 5 & 30 \end{bmatrix}$$

$$\mathbf{y}^t\mathbf{x} = [1 \quad 6]\begin{bmatrix} 3 \\ 5 \end{bmatrix} = [3+30] = [33] = 33,$$

since we will drop the brackets from 1×1 matrices. □

Remark. The vector product $\mathbf{x}\mathbf{y}^t$ is a matrix called the *outer product* of $\mathbf{x}$ and $\mathbf{y}$. Decompositions of large matrices into sums of outer products are an important tool in modern image analysis. □

The product BA does not have to be defined even if AB is defined.

Example 5. Let $A = \begin{bmatrix} 1 & 4 \\ 2 & 0 \\ 5 & 1 \end{bmatrix}$ and $B = \begin{bmatrix} 7 & -1 \\ 0 & -3 \end{bmatrix}$, then

$$AB = \begin{bmatrix} 1 & 4 \\ 2 & 0 \\ 5 & 1 \end{bmatrix} = \left[A\begin{bmatrix} 7 \\ 0 \end{bmatrix} \;\; A\begin{bmatrix} -1 \\ -3 \end{bmatrix} \right] = \begin{bmatrix} 7 & -13 \\ 14 & -2 \\ 35 & -8 \end{bmatrix}.$$

But, BA is not defined since the product of B with a column vector of length 3 is not defined. □

In spite of the fact that in general the factors in a matrix product cannot be rearranged without changing the product, matrix multiplication is algebraically well behaved and most of the familiar laws of arithmetic hold. They are summarized below in Theorem 1.

There is a special $n\times n$ matrix which when multiplied by an $n\times n$ matrix B just gives B itself as the result.

Definition. The $(n\times n)$ *identity matrix* I is the $n\times n$ matrix which satisfies

$$I\mathbf{x} = \mathbf{x} \textit{ for all } \mathbf{x} \text{ in } \mathbb{R}^n.$$

In view of Theorem 1 of the previous section and the fact that $I\mathbf{e}_j = \mathbf{e}_j$ must hold for $j = 1,2,\ldots,n$, we see that the identity matrix satisfies

$$\begin{aligned} I(k,k) &= 1 \qquad k = 1,2,\ldots,n \\ I(j,k) &= 0 \qquad \textit{if } j \neq k,\ 1 \leq j,k \leq n. \end{aligned}$$

The 2×2 and 3×3 identity matrices are

$$\begin{bmatrix} 1 & 0 \\ 0 & 1 \end{bmatrix} \text{ and } \begin{bmatrix} 1 & 0 & 0 \\ 0 & 1 & 0 \\ 0 & 0 & 1 \end{bmatrix}.$$

Theorem 1. Let A, B, and C be matrices of compatible sizes, and let s be a real number, then

$$\begin{array}{lrl} \text{(i)} & A(B+C) & = AB + AC \\ \text{(ii)} & (A+B)C & = AC + BC \\ \text{(iii)} & s(AB) & = (sA)B = A(sB) \\ \text{(iv)} & A(BC) & = (AB)C \\ \text{(v)} & IA & = A. \end{array}$$

Remark. This list implies properties of the inner product. For example, if **x**, **y**, and **z** are in $\mathbb{R}^n$, then (i) says $\mathbf{x}^t(\mathbf{y} + \mathbf{z}) = \mathbf{x}^t\mathbf{y} + \mathbf{x}^t\mathbf{z}$.

2.2 EXERCISES

1. Find all possible products of pairs of the following matrices

$$A = [1 \quad 2 \quad 4 \quad 6] \qquad B = \begin{bmatrix} 2 & 3 & 5 \\ -1 & 0 & 1 \\ 4 & 0 & 0 \end{bmatrix} \qquad C = \begin{bmatrix} 4 & 6 \\ 7 & 8 \\ 0 & 5 \end{bmatrix}$$

$$D = \begin{bmatrix} 3 & 3 \\ 1 & 2 \\ 2 & 0 \\ 3 & 4 \end{bmatrix} \qquad E = \begin{bmatrix} 3 & 7 \\ -4 & 1 \end{bmatrix}$$

2. Let $R = \begin{bmatrix} 0 & -1 \\ 1 & 0 \end{bmatrix}$. Verify that $RR^t = R^tR = I$.

3. Let $S = \begin{bmatrix} 0 & 1 \\ 1 & 0 \end{bmatrix}$. Verify that $SS = I$.

4. Let $M = R^tSR$. What is MM? Use problems 2 and 3, no computations are needed.

5. Let $H = \begin{bmatrix} 1 & -1 \\ 1 & -1 \end{bmatrix}$. Find HH.

2.3. Inverses of Matrices

In this section, we introduce the concept of the inverse of a matrix and discuss its theoretical uses and practical limitations in the solution of linear systems.

Consider the linear system $A\mathbf{x} = \mathbf{b}$. Suppose that for any choice of $\mathbf{b}$ there is a *unique* solution vector $\mathbf{x}$. This forces A to be square. Intuitively, the only systems that can have unique solutions for *every* possible right hand side are those systems having exactly as many equations as unknowns. Overdetermined systems (more equations than unknowns) require very special $\mathbf{b}$ vectors and underdetermined systems with $\mathbf{b} = \mathbf{0}$ have infinitely many solutions. In Chapter 3 we develop these ideas in more detail. The following fact states that the operation that takes the right hand side $\mathbf{b}$ to the solution vector $\mathbf{x}$ is well behaved.

Theorem 1. Let A be an $n \times n$ matrix with the property that the system

$$A\mathbf{x} = \mathbf{b}$$

has a unique solution for every given $\mathbf{b}$. Then there is an $n \times n$ matrix called the *inverse* of A and denoted by A^{-1} so that

$$\mathbf{x} = A^{-1}\mathbf{b}.$$

Definition. An $n \times n$ matrix that has an inverse is called *invertible* or *non-singular.*

Example 1. Verify that the solution of

$$\begin{bmatrix} a & b \\ c & d \end{bmatrix} \begin{bmatrix} x \\ y \end{bmatrix} = \begin{bmatrix} \alpha \\ \beta \end{bmatrix}$$

is

$$\begin{bmatrix} x \\ y \end{bmatrix} = \begin{bmatrix} \frac{d}{\Delta} & -\frac{b}{\Delta} \\ -\frac{c}{\Delta} & \frac{a}{\Delta} \end{bmatrix} \begin{bmatrix} \alpha \\ \beta \end{bmatrix} \tag{1}$$

where $\Delta = ad - bc \neq 0$.

Solution. We have to show

$$\begin{aligned} ax + by &= \alpha \\ cx + dy &= \beta \end{aligned}$$

if $x = \dfrac{\alpha d - b\beta}{\Delta}$ and $y = \dfrac{a\beta - c\alpha}{\Delta}$. So, we do the algebra

$$a\left\{\frac{\alpha d - b\beta}{\Delta}\right\} + b\left\{\frac{a\beta - c\alpha}{\Delta}\right\} = \frac{a\alpha d - ab\beta - \alpha bc + a\beta b}{\Delta} = \frac{\alpha(ad-bc)}{\Delta} = \alpha.$$

The second equation is similar. □

We could have derived formula (1) by algebraic manipulation. There is a general strategy for computing inverses that also gives (1). The strategy relies on the fact that if M is an $n \times n$ matrix, then

$$M = [M\mathbf{e}_1 \quad M\mathbf{e}_2 \quad \cdots \quad M\mathbf{e}_n]$$

where $\mathbf{e}_1, \mathbf{e}_2, \ldots, \mathbf{e}_n$ are the standard unit vectors in $\mathbb{R}^n$ (see Section 2.1). Now, taking M to be A^{-1} we see that we must find

$$A^{-1}\mathbf{e}_1,\ A^{-1}\mathbf{e}_2, \ldots, A^{-1}\mathbf{e}_n.$$

Now,

$$A^{-1}\mathbf{e}_i = \mathbf{c}_i \quad \textit{if and only if} \quad A\mathbf{c}_i = \mathbf{e}_i.$$

This shows

***Theorem* 2.** If A is an $n \times n$ invertible matrix, then the columns of A^{-1} are $\mathbf{c}_1, \mathbf{c}_2, \ldots, \mathbf{c}_n$ where $A\mathbf{c}_i = \mathbf{e}_i$, $i = 1, 2, \ldots, n$.

***Example* 2.** Use Theorem 2 to verify formula (1).

Solution. We are to solve

$$ax + by = 1 \qquad\qquad au + bv = 0$$

and

$$cx + dy = 0 \qquad\qquad cu + dv = 1$$

We will assume $a \neq 0$ and use Gaussian Elimination (if $a = 0$ we interchange rows). This gives the upper triangular systems

$$\begin{aligned} ax + by &= 1 \\ (d - \tfrac{bc}{a})y &= -\tfrac{c}{a} \end{aligned} \qquad\qquad \begin{aligned} au + bv &= 0 \\ (d - \tfrac{bc}{a})v &= 1. \end{aligned}$$

Backsolving gives

$$y = \frac{-c}{ad - bc} \qquad\qquad v = \frac{a}{ad - bc}$$

$$x = \frac{d}{ad - bc} \qquad\qquad u = \frac{-b}{ad - bc}$$

as required. □

Example 3. Find the inverse of $A = \begin{bmatrix} 0 & 1 & 2 \\ 0 & 2 & 3 \\ 1 & 1 & 1 \end{bmatrix}$, see Example 4 of Section 1.3.

Solution. In Example 4 of Section 1.3 we found

$$A\begin{bmatrix} 1 \\ -3 \\ 2 \end{bmatrix} = \begin{bmatrix} 1 \\ 0 \\ 0 \end{bmatrix} \text{ and } A\begin{bmatrix} -1 \\ 2 \\ -1 \end{bmatrix} = \begin{bmatrix} 0 \\ 1 \\ 0 \end{bmatrix}.$$

So, we must solve $A\mathsf{x} = \mathsf{e}_3$. Gaussian Elimination gives the system (see equation (7) of Example 4, Section 1.3)

$$\begin{aligned} x_2 + 2x_3 &= 0 \\ x_1 + x_2 + x_3 &= 1 \end{aligned}$$

which has the solution $x_1 = 1$, $x_2 = 0$, $x_3 = 0$. Therefore,

$$A^{-1} = \begin{bmatrix} 1 & -1 & 1 \\ -3 & 2 & 0 \\ 2 & -1 & 0 \end{bmatrix}. \qquad \square$$

The formula $\mathsf{x} = A^{-1}\mathsf{b}$ shows

$$\mathsf{b} = A\mathsf{x} = A(A^{-1}\mathsf{b}) = (AA^{-1})\mathsf{b}$$

for all b. Therefore AA^{-1} must be the identity matrix:

$$AA^{-1} = I \tag{2}$$

Similarly,

$$\mathsf{x} = A^{-1}\mathsf{b} = A^{-1}(A\mathsf{x}) = (A^{-1}A)\mathsf{x}$$

gives

$$A^{-1}A = I \tag{3}$$

Example 4. Verify (2) and (3) for the matrix of Example 3.

Solution. For (2)

$$\begin{bmatrix} 0 & 1 & 2 \\ 0 & 2 & 3 \\ 1 & 1 & 1 \end{bmatrix}\begin{bmatrix} 1 & -1 & 1 \\ -3 & 2 & 0 \\ 2 & -1 & 0 \end{bmatrix} = \begin{bmatrix} 4-3 & 2-2 & 0 \\ -6+6 & 4-3 & 0 \\ 1-3+2 & -1+2-1 & 1 \end{bmatrix} = I$$

For (3)

$$\begin{bmatrix} 1 & -1 & 1 \\ -3 & 2 & 0 \\ 2 & -1 & 0 \end{bmatrix}\begin{bmatrix} 0 & 1 & 2 \\ 0 & 2 & 3 \\ 1 & 1 & 1 \end{bmatrix} = \begin{bmatrix} 1 & 1-2+1 & 2-3+1 \\ 0 & -3+4 & -6+6 \\ 0 & 2-2 & 4-3 \end{bmatrix} = I \qquad \square$$

If A and B are invertible matrices, then the inverse of AB is $B^{-1}A^{-1}$. That is

$$(AB)^{-1} = B^{-1}A^{-1}.$$

To see this, suppose that $B\mathsf{x} = \mathsf{z}$ and $A\mathsf{z} = \mathsf{d}$ so $(AB)\mathsf{x} = \mathsf{d}$. Then, symbolically, $\mathsf{x} = (AB)^{-1}\mathsf{d}$. But also, $\mathsf{x} = B^{-1}\mathsf{z}$ and $\mathsf{z} = A^{-1}\mathsf{d}$ so $\mathsf{x} = B^{-1}A^{-1}\mathsf{d}$. So, the symbol $(AB)^{-1}$ must be the product $B^{-1}A^{-1}$. This is sometimes called the "Shoes and Socks Theorem." To undo a series of actions, the last action must be the first one undone. If we apply A to the result of B, then we must first undo the effect of A by multiplying by A^{-1}. We then undo the effect of B.

The inverse has ***no computational advantage*** over Gaussian Elimination for solving linear systems. The work spent in finding the inverse is equivalent to performing Gaussian Elimination several times; so even if you save your multipliers, as advocated in Section 1.3, you pay a lot up front to get the inverse. In addition, the effort spent in forming the product $A^{-1}\mathsf{b}$ is the same as that spent in preprocessing b and then backsolving. So, after you've invested time and effort to find A^{-1}, you still have no advantage. In fact in some systems you have a distinct disadvantage.

Example 5. Let

$$A = \begin{bmatrix} 2 & -1 & 0 & 0 & 0 \\ -1 & 2 & -1 & 0 & 0 \\ 0 & -1 & 2 & -1 & 0 \\ 0 & 0 & -1 & 2 & -1 \\ 0 & 0 & 0 & -1 & 2 \end{bmatrix}$$

Verify that

$$A^{-1} = \frac{1}{6}\begin{bmatrix} 5 & 4 & 3 & 2 & 1 \\ 4 & 8 & 6 & 4 & 2 \\ 3 & 6 & 9 & 6 & 3 \\ 2 & 4 & 6 & 8 & 4 \\ 1 & 2 & 3 & 4 & 5 \end{bmatrix},$$

and show that the solution of $A\mathsf{x} = \mathsf{b}$ by Gaussian Elimination takes fewer operations than application of the formula $\mathsf{x} = A^{-1}\mathsf{b}$.

Solution. The verification $AA^{-1} = I$ is a direct computation. Now, to form $A^{-1}\mathsf{b}$ requires 5 multiplications, 4 additions, and 1 division (by 6) for each coordinate. So the total work — for general b — is

25 *multiplications*, 20 *additions*, 5 *divisions.*

In Gaussian Elimination the elimination of x_1 from the second equation first requires that we find the multiplier $-\frac{1}{2}$. This takes one division. We then multiply the coefficient x_2 in the first equation by this number and subtract the result from the coefficient of x_2 in the second equation. We do a similar operation on the right hand side. So, the processing of the second equation takes 1 division, 2 multiplications, and 2 additions. Since four equations are to be processed we use totals of

$$8 \textit{ multiplications, } 4 \textit{ divisions, and } 8 \textit{ additions.}$$

Backsolving takes a total of 5 divisions and 4 additions.

If we count divisions as multiplications, we have

	Multiplications	Additions
Gaussian Elimination	17	12
Application of A^{-1}	30	20.

Finally, recall that these figures don't include the computations needed to form A^{-1}! □

2.3. EXERCISES

1. Example 1 shows that

$$\begin{bmatrix} a & b \\ c & d \end{bmatrix}^{-1} = \begin{bmatrix} \frac{d}{\Delta} & -\frac{b}{\Delta} \\ -\frac{c}{\Delta} & \frac{a}{\Delta} \end{bmatrix}$$

where $\Delta = ad - bc \neq 0$. Use this to find the inverses of the following matrices.

(a) $\begin{bmatrix} 2 & 1 \\ -2 & 2 \end{bmatrix}$ (b) $\begin{bmatrix} 0 & -1 \\ 1 & 0 \end{bmatrix}$ (c) $\begin{bmatrix} 8 & 5 \\ 1 & 3 \end{bmatrix}$

(d) $\begin{bmatrix} 0 & 1 \\ 1 & 0 \end{bmatrix}$ (e) $\begin{bmatrix} 4 & 0 \\ 0 & 3 \end{bmatrix}$ (f) $\begin{bmatrix} -1 & -1 \\ -1 & 0 \end{bmatrix}$

2. Show that the matrices below do not have inverses

(a) $\begin{bmatrix} 1 & 1 \\ -1 & -1 \end{bmatrix}$ (b) $\begin{bmatrix} 0 & 3 \\ 0 & 0 \end{bmatrix}$ (c) $\begin{bmatrix} 1 & 2 \\ 1 & 2 \end{bmatrix}$

3. Find the inverse of

$$\begin{bmatrix} 1.2969 & 0.8648 \\ 0.2161 & 0.1441 \end{bmatrix}.$$

(See Problem 5 of Section 1.2.)

4. (a) Convince yourself that if A is a 3×3 matrix, $A = [\mathsf{u}_1 \quad \mathsf{u}_2 \quad \mathsf{u}_3]$, with

$$\mathsf{u}_i^t \mathsf{u}_j = \begin{cases} 0 \text{ if } i \neq j \\ 1 \text{ if } i = j \end{cases}$$

then $A^{-1} = A^t$.

(b) Verify that the matrices below have the property described in part (a). Then, verify that the inverse of each is its transpose.

(i) $\begin{bmatrix} \frac{1}{\sqrt{2}} & 0 & -\frac{1}{\sqrt{2}} \\ 0 & 1 & 0 \\ \frac{1}{\sqrt{2}} & 0 & \frac{1}{\sqrt{2}} \end{bmatrix}$ (ii) $\begin{bmatrix} 0 & 0 & 1 \\ 0 & 1 & 0 \\ 1 & 0 & 0 \end{bmatrix}$

(iii) $\frac{1}{\sqrt{6}} \begin{bmatrix} \sqrt{2} & 1 & \sqrt{3} \\ \sqrt{2} & -2 & 0 \\ \sqrt{2} & 1 & -\sqrt{3} \end{bmatrix}$ (iv) $\begin{bmatrix} 0 & 1 & 0 \\ 0 & 0 & 1 \\ 1 & 0 & 0 \end{bmatrix}$

5. Verify the Shoes and Socks Theorem by showing $(AB)^{-1} = B^{-1}A^{-1}$ for some pairs A, B chosen from Problem 1.

CHAPTER 3 Geometric Theory of Linear Systems

In this chapter we study the geometric theory of small linear systems and begin to learn the basic concepts of Linear Algebra. Gaussian Elimination provides an algebraic way of determining the solution set of general $n \times k$ linear system $A\mathbf{x} = \mathbf{b}$. However, in many problems that arise in scientific and statistical applications one deals with systems having either no solution or infinitely many solutions. Gaussian Elimination is not the appropriate tool for these systems. We will study a geometric approach that gives another perspective on linear systems and will enable us to find solutions to these problems in some cases.

3.1. The Column Space of a Matrix

The matrix-vector product $A\mathbf{x}$ is defined algebraically in terms of the rows of A: the ith entry of $A\mathbf{x}$ is the inner product of $\mathbf{x}$ with the ith row of A. It is useful to see how the columns of A influence the product $A\mathbf{x}$. There are no hard computations; you just have to keep your eyes opened. The following example illustrates the general idea.

$$\begin{bmatrix} 1 & 2 \\ 3 & 4 \\ 5 & 6 \end{bmatrix} \begin{bmatrix} x \\ \\ y \end{bmatrix} = \begin{bmatrix} x+2y \\ 3x+4y \\ 5x+6y \end{bmatrix} = \begin{bmatrix} x \\ 3x \\ 5x \end{bmatrix} + \begin{bmatrix} 2y \\ 4y \\ 6y \end{bmatrix} = x\begin{bmatrix} 1 \\ 3 \\ 5 \end{bmatrix} + y\begin{bmatrix} 2 \\ 4 \\ 6 \end{bmatrix}.$$

The first equality is just the definition of the matrix-vector product. The second is a consequence of the definition of addition of vectors; we used it to separate components rather than to combine them. Finally, the third equality is a consequence of the rule for scaling a vector by a real number — the common factors of x and y were taken out.

Now, observe that the column vectors

$$\begin{bmatrix} 1 \\ 3 \\ 5 \end{bmatrix} \text{ and } \begin{bmatrix} 2 \\ 4 \\ 6 \end{bmatrix}$$

are the columns of our original matrix. This is no accident. We have the following general principle.

***Theorem* 1.** If A is an $n \times k$ matrix whose column vectors (from left to right) are $\mathbf{c}_1, \mathbf{c}_2, \ldots, \mathbf{c}_k$ and if $\mathbf{x}$ is a column vector of length k, then

$$A\mathbf{x} = x_1\mathbf{c}_1 + x_2\mathbf{c}_2 + \cdots + x_k\mathbf{c}_k.$$

Example 1. Let $A = \begin{bmatrix} 1 & 8 & 0 \\ 2 & 9 & 0 \\ 3 & 0 & 4 \end{bmatrix}$, then with $\mathbf{x} = \begin{bmatrix} x \\ y \\ z \end{bmatrix}$ $A\mathbf{x}$ can be viewed as either

$$x\begin{bmatrix} 1 \\ 2 \\ 3 \end{bmatrix} + y\begin{bmatrix} 8 \\ 9 \\ 0 \end{bmatrix} + z\begin{bmatrix} 0 \\ 0 \\ 4 \end{bmatrix} \text{ or } \begin{bmatrix} x+8y \\ 2x+9y \\ 3x+4z \end{bmatrix}$$

The power of Theorem 1 is that it interprets an algebraic entity, the matrix-vector product, in geometric terms, as a weighted sum of vectors. Expressions of the form $s_1\mathbf{v}_1 + s_2\mathbf{v}_2 + \ldots + s_j\mathbf{v}_j$ occur quite often so they deserve a special name.

Definition. A *linear combination* of the vectors $\mathbf{v}_1,\mathbf{v}_2,\ldots,\mathbf{v}_j$ is any expression of the form $s_1\mathbf{v}_1 + s_2\mathbf{v}_2 + \ldots + s_j\mathbf{v}_j$ where $s_1,s_2,\ldots,s_j$ are real numbers.

From Theorem 1 we can see that the set of all "outputs" of a given matrix is exactly the set of all linear combinations of the column vectors.

Definition. Let A be an $n \times k$ matrix. The set of all linear combinations of the columns of A is called the *column space of A*.

The column space of A is also called the *range of A*. The relationship between the column space of A and the solvability of the system $A\mathbf{x} = \mathbf{b}$ is summarized in

Theorem 2. The system $A\mathbf{x} = \mathbf{b}$ has a solution if and only if $\mathbf{b}$ is in the column space of A.

In determining the solvability of $A\mathbf{x} = \mathbf{b}$ when A has two or three rows, we can think of the columns of A as pointing in "permissible directions of travel" in two or three dimensional space. The system $A\mathbf{x} = \mathbf{b}$ has a solution only if $\mathbf{b}$ can be reached from the origin by travelling in these directions and their opposites. If the columns of A are $\mathbf{a}_1,\ldots,\mathbf{a}_k$ and if we reach $\mathbf{b}$ after proceeding along the vectors $s_1\mathbf{a}_1, s_2\mathbf{a}_2, \ldots,$ and $s_k\mathbf{a}_k$, then $\mathbf{b}$ must be the sum $s_1\mathbf{a}_1 + s_2\mathbf{a}_2 + \cdots + s_k\mathbf{a}_k$.

Example 2. Let $A = \begin{bmatrix} 1 & 2 \\ 2 & 4 \end{bmatrix}$. Any vector of the form

$$x\begin{bmatrix} 1 \\ 2 \end{bmatrix} + y\begin{bmatrix} 2 \\ 4 \end{bmatrix} \tag{1}$$

is a linear combination of the columns of A.

The set of all linear combinations of columns of A is the set of all vectors that can be written in the form (1). We will analyze this set algebraically and geometrically. First, algebraically we have

$$x\begin{bmatrix}1\\2\end{bmatrix} + y\begin{bmatrix}2\\4\end{bmatrix} = x\begin{bmatrix}1\\2\end{bmatrix} + 2y\begin{bmatrix}1\\2\end{bmatrix} = (x+2y)\begin{bmatrix}1\\2\end{bmatrix} \tag{2}$$

which says that any linear combination of the columns of A is just a multiple of the first column.

Geometrically, we can see that if we add a multiple of $\begin{bmatrix}1\\2\end{bmatrix}$ to a multiple of $\begin{bmatrix}2\\4\end{bmatrix}$, we end up with a vector which is still a multiple of $\begin{bmatrix}1\\2\end{bmatrix}$. That is, we never leave the line that passes through the origin heading in the direction of $\begin{bmatrix}1\\2\end{bmatrix}$.

From these observations we conclude that the column space of A is the straight line through the origin with slope 2. Finally, we note that the system $A\mathbf{x} = \mathbf{b}$ has a solution if and only if $\mathbf{b}$ lies on this line. That is, $A\mathbf{x} = \mathbf{b}$ has a solution if and only if $b_2 = 2b_1$.

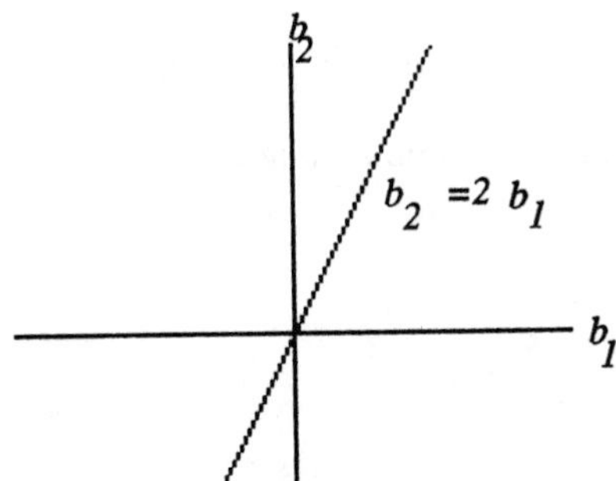

Figure 1. The column space of A is the line $b_2 = 2b_1$. □

If the column vectors of a two by two matrix point in different directions, then the column space must be all of $\mathbb{R}^2$. You can see this geometrically. If $\mathbf{u}$ and $\mathbf{v}$ are vectors in the plane which make an angle strictly between 0 and 180 degrees, then any point in the plane can be reached from the origin by following only the directions of $\mathbf{u}$ and $\mathbf{v}$ and their negatives. That is, any vector can be realized as a multiple of $\mathbf{u}$ plus a multiple of $\mathbf{v}$. If the column vectors are multiples of each other, then the column space is simply the line through the origin in the direction of the columns.

***Theorem* 3.** If A is a $2\times k$ matrix, then the column space is either $\mathbb{R}^2$, a line in $\mathbb{R}^2$ through the origin, or the zero vector by itself.

Only for the zero matrix is the column space the zero vector.

***Example* 3.** Find the column space of each of the following matrices.

$$A = \begin{bmatrix} 1 & 1 & 2 & 3 \\ 0 & 0 & 1 & 0 \end{bmatrix} \quad B = \begin{bmatrix} 0 & 0 & 1 \\ 0 & 0 & 5 \end{bmatrix} \quad C = \begin{bmatrix} 3 & 5 \\ 2 & -1 \end{bmatrix} \quad D = \begin{bmatrix} 2 & -1 \\ 4 & -2 \end{bmatrix}$$

Solution. The first and third columns of A are not multiples of each other so the column space of A is all of $\mathbb{R}^2$. The system $A\mathbf{x} = \mathbf{b}$ always has a solution.

The column space of B is the set of all multiples of the third column. This is the straight line through the origin with slope 5 — the line whose equation is $y = 5x$. The system $B\mathbf{x} = \mathbf{b}$ has a solution if and only if $b_2 = 5b_1$.

The columns of C are not multiples of each other so the column space is $\mathbb{R}^2$. The system $C\mathbf{x} = \mathbf{b}$ has a solution for every $\mathbf{b}$.

In the matrix D the second column is $-\frac{1}{2}$ times the first so the column space is the set of all multiples of the first column. This is the line through the origin with slope 2. The system $D\mathbf{x} = \mathbf{b}$ has a solution if and only if $b_2 = 2b_1$.

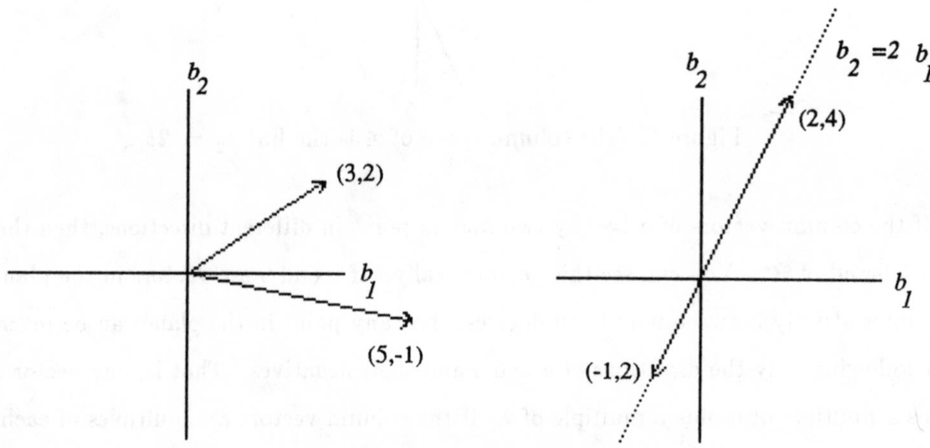

Figure 2. The Column Vectors of Matrices C and D. □

For matrices with three rows the column space must be a subset of $\mathbb{R}^3$.

Theorem 4. Let A be a matrix with exactly three rows. The column space of A is one of the following: the zero vector, a line through the origin, a plane containing the origin, or all of $\mathbb{R}^3$. The system $A\mathbf{x} = \mathbf{b}$ has a solution for every choice of $\mathbf{b}$ if and only if the column space is all of $\mathbb{R}^3$.

The following examples illustrate these possibilities.

Example 4. For what vectors $\mathbf{b}$ does the system $A\mathbf{x} = \mathbf{b}$ have a solution if

$$A = \begin{bmatrix} 1 & 2 \\ 0 & 1 \\ 1 & 1 \end{bmatrix}?$$

Solution. Since the columns are not multiples of each other, the column space is not a line. Since two vectors (or three points) determine a plane in $\mathbb{R}^3$, we see that the column space is a plane. The parametric equations of this plane are

$$s\begin{bmatrix} 1 \\ 0 \\ 1 \end{bmatrix} + t\begin{bmatrix} 2 \\ 1 \\ 1 \end{bmatrix}.$$

The algebraic equation of the plane is much easier to use in checking membership in the plane. Since the normal to the plane is $(1,0,1)\times(2,1,1) = (-1,1,1)$, this equation is

$$x - y - z = 0.$$

The system $A\mathbf{x} = \mathbf{b}$ has a solution if and only if $\mathbf{b}$ lies in this plane; that is, $b_1 - b_2 - b_3 = 0$.

For example,

$$\begin{aligned} x + 2y &= 3 \\ y &= 4 \\ x + y &= 10 \end{aligned}$$

has no solution since $3 - 4 - 10 \neq 0$.

On the other hand the system

$$\begin{aligned} x + 2y &= 3 \\ y &= 1 \\ x + y &= 2 \end{aligned} \tag{3}$$

has the unique solution $x = y = 1$. □

Example 5. For what values of a, b, c does the system

$$\begin{aligned} x + 2y + 3z &= a \\ 2x + 4y + 6z &= b \\ -3x - 6y - 9z &= c \end{aligned}$$

have a solution?

Solution. The columns of the coefficient matrix are all multiples of $\begin{bmatrix} 1 \\ 2 \\ -3 \end{bmatrix}$. Therefore, the system has a solution if and only if

$$\begin{bmatrix} a \\ b \\ c \end{bmatrix} = t \begin{bmatrix} 1 \\ 2 \\ -3 \end{bmatrix}$$

holds for some number t. That is, $b = 2a$, $c = -3a$ must hold. Geometrically, the column space is the line through the origin in the direction of $\begin{bmatrix} 1 \\ 2 \\ -3 \end{bmatrix}$. □

Example 6. Let $A = \begin{bmatrix} 1 & 2 & -1 \\ 0 & 1 & -2 \\ 1 & -1 & 5 \end{bmatrix}$ and consider the following choices of b.

$$\mathsf{b} = \begin{bmatrix} 1 \\ 1 \\ -2 \end{bmatrix}, \begin{bmatrix} -2 \\ 1 \\ 1 \end{bmatrix}, \begin{bmatrix} 1 \\ 2 \\ 1 \end{bmatrix} \text{ and } \begin{bmatrix} 0 \\ 1 \\ 1 \end{bmatrix}.$$

For which b does $A\mathsf{x} = \mathsf{b}$ have a solution?

Solution. We will find the column space of A and determine which, if any, of the b's are in it. First notice that the first two columns are not multiples of each other, so they determine a plane through the origin. The normal vector to this plane is the cross product of the first two columns of A

$$\mathsf{n} = \begin{bmatrix} 1 \\ 0 \\ 1 \end{bmatrix} \times \begin{bmatrix} 2 \\ 1 \\ -1 \end{bmatrix} = \begin{bmatrix} -1 \\ 3 \\ 1 \end{bmatrix}.$$

Therefore, the equation of this plane is $-x + 3y + z = 0$. Now if the remaining column of A is in this plane, then the range of A must be this plane. If the third column is not in this plane, then it is independent of the two vectors defining this plane and the range of A will have to be all of $\mathbb{R}^3$. We check if the coordinates of the third column satisfy $-x + 3y + z = 0$:

$$-(-1) + 3(-2) + 5 = 0.$$

Therefore, the range of A is the plane $-x + 3y + z = 0$.

Now, we should check each of the given b vectors to see if its coordinates satisfy this equation. Of the given vectors only the first one $\begin{bmatrix} 1 \\ 1 \\ -2 \end{bmatrix}$ is in the range of A.

So, for example, the system

$$\begin{aligned} x + 2y - z &= 0 \\ y - 2z &= 1 \\ x - y + 5z &= 1 \end{aligned} \qquad (4)$$

has no solution. □

Remark. The problem of finding the range of a general $n \times k$ matrix is quite difficult in general. For example, if the matrix has four or more rows, there is no good tool like the cross product. However, the purpose here is ***not*** to provide tools for finding column spaces but rather to introduce the reader to the concept through concrete geometrical examples. □

Example 7. Show that the system
$$\begin{bmatrix} 1 & 1 & 0 \\ 3 & 4 & 1 \\ 0 & 1 & 5 \end{bmatrix} \begin{bmatrix} x \\ y \\ z \end{bmatrix} = \begin{bmatrix} a \\ b \\ c \end{bmatrix}$$

has a solution for every choice of a, b, c.

Solution. We will show that the column space is all of $\mathbb{R}^3$. The first two columns determine a plane whose normal vector is

$$\mathbf{n} = \begin{bmatrix} 1 \\ 3 \\ 0 \end{bmatrix} \times \begin{bmatrix} 1 \\ 4 \\ 1 \end{bmatrix} = \begin{bmatrix} 3 \\ -1 \\ 1 \end{bmatrix}$$

and whose algebraic equation is

$$3x - y + z = 0.$$

The components of the third column do not satisfy this equation:

$$3(0) - 1 + 5 \neq 0.$$

Geometrically, this means that the plane $3x - y + z = 0$ and the third column vector are configured something like this:

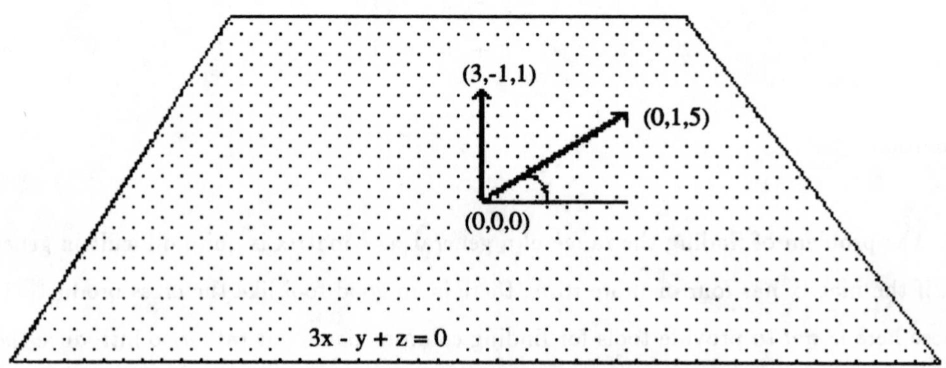

From this it follows that any point in $\mathbb{R}^3$ can be reached from the origin by first moving in the plane $3x - y + z = 0$ and then moving parallel to $\begin{bmatrix} 0 \\ 1 \\ 5 \end{bmatrix}$. Algebraically, this means that every point in $\mathbb{R}^3$ is a linear combination of the columns of A. In fact, for a given $\mathbf{b}$ in $\mathbb{R}^3$ there is only one linear combination that will do. One way to see this is to note that there is exactly one point in the plane $3x - y + z = 0$ from which we can get to $\mathbf{b}$ by following the direction of $\begin{bmatrix} 0 \\ 1 \\ 5 \end{bmatrix}$, and once we follow this direction the distance we follow it is uniquely determined. □

Dimension

At the intuitive level the dimension of a set is the minimal number of parameters needed to completely specify the set. At present, we are interested mainly in dimensions of Euclidean spaces and of column spaces of matrices. The set $\mathbb{R}^n$ has dimension n because members of $\mathbb{R}^n$ are uniquely determined by their n coordinates. Points have dimension 0 by convention; lines have dimension 1; and planes have dimension 2.

The following example further illustrates the concept of dimension.

Example 8. Find the dimension of the column space of

$$M = \begin{bmatrix} 1 & 1 & 1 & 0 & 0 \\ 0 & 1 & 0 & 1 & 0 \\ 1 & 0 & 0 & 0 & 1 \\ 1 & 0 & 0 & 0 & 1 \end{bmatrix}.$$

Solution. The column space is a subset of $\mathbb{R}^4$, so its dimension must be no greater than four. Since there are five columns, there must be some redundancy. By inspection, we see that the second column is the sum of the third and fourth:

$$\begin{bmatrix} 1 \\ 1 \\ 0 \\ 0 \end{bmatrix} = \begin{bmatrix} 1 \\ 0 \\ 0 \\ 0 \end{bmatrix} + \begin{bmatrix} 0 \\ 1 \\ 0 \\ 0 \end{bmatrix}.$$

Also, the first column is the sum of the third and fifth:

$$\begin{bmatrix} 1 \\ 0 \\ 1 \\ 1 \end{bmatrix} = \begin{bmatrix} 1 \\ 0 \\ 0 \\ 0 \end{bmatrix} + \begin{bmatrix} 0 \\ 0 \\ 1 \\ 1 \end{bmatrix}.$$

So, any member of the column space can be expressed in terms of only

$$\begin{bmatrix} 1 \\ 0 \\ 0 \\ 0 \end{bmatrix}, \begin{bmatrix} 0 \\ 1 \\ 0 \\ 0 \end{bmatrix}, \text{ and } \begin{bmatrix} 0 \\ 0 \\ 1 \\ 1 \end{bmatrix},$$

hence, the dimension of the column space is three. □

3.1. EXERCISES

1. Verify that the unique solution of system (3) in Example 4 is $x = y = 1$.

2. Verify, with Gaussian Elimination, that system (4) in Example 6 has no solution.

3. Use Gaussian Elimination to show that the system in Example 7 has a unique solution for every choice of a, b, and c.

4. Find the column space of each of the following matrices.

(a) $\begin{bmatrix} 2 & 3 \\ 1 & 4 \end{bmatrix}$ (b) $\begin{bmatrix} 0 & 1 \\ 1 & 0 \end{bmatrix}$ (c) $\begin{bmatrix} 4 & 5 \\ 8 & 10 \end{bmatrix}$ (d) $\begin{bmatrix} -1 & 2 \\ 3 & -6 \end{bmatrix}$

(e) $\begin{bmatrix} 1 & 2 & -1 \\ 0 & 1 & -2 \\ 1 & -1 & 5 \end{bmatrix}$ (f) $\begin{bmatrix} 1 & 0 & 0 \\ 2 & 2 & 3 \\ 0 & 1 & -1 \end{bmatrix}$ (g) $\begin{bmatrix} 2 & 2 \\ 1 & 1 \\ 3 & 3 \end{bmatrix}$

(h) $\begin{bmatrix} 1 & 2 & 3 & 4 \\ 5 & 6 & 7 & 8 \\ 9 & 10 & 11 & 12 \end{bmatrix}$

5. The ranges of the following matrices are planes. In each case (a) find the equation of the plane, (b) determine if the system $A\mathbf{x} = \mathbf{b}$ has a solution for the given vectors $\mathbf{b}$ and (c) if there is a solution, find all solutions.

(i) $\begin{bmatrix} 1 & 4 & 6 \\ 2 & 1 & 5 \\ 1 & -1 & 1 \end{bmatrix}$ $\quad \mathbf{b} = \begin{bmatrix} 2 \\ 2 \\ 1 \end{bmatrix}$ (ii) $\begin{bmatrix} 1 & 4 & 5 \\ 2 & 2 & 4 \\ 2 & -1 & 1 \end{bmatrix}$ $\quad \mathbf{b} = \begin{bmatrix} 1 \\ 0 \\ 2 \end{bmatrix}$

(iii) $\begin{bmatrix} 1 & 4 \\ 2 & 5 \\ 0 & 1 \end{bmatrix}$ $\quad \mathbf{b} = \begin{bmatrix} 8 \\ 0 \\ 2 \end{bmatrix}$ and $\mathbf{b} = \begin{bmatrix} 2 \\ 1 \\ 1 \end{bmatrix}$

(iv) $\begin{bmatrix} 2 & 4 & 6 \\ 0 & 1 & -1 \\ 3 & 7 & 8 \end{bmatrix}$ $\quad \mathbf{b} = \begin{bmatrix} 2 \\ 0 \\ 3 \end{bmatrix}$ and $\mathbf{b} = \begin{bmatrix} 3 \\ 4 \\ 5 \end{bmatrix}$.

6. Show that the dimension of the column space is four for each matrix below

$$\begin{bmatrix} 1 & 1 & 1 & 0 & 0 & 0 \\ 0 & 0 & 0 & 0 & 0 & 0 \\ 1 & 2 & 0 & 1 & 0 & 0 \\ 0 & 0 & 0 & 0 & 0 & 0 \\ 1 & 0 & 0 & 0 & 2 & 0 \\ 0 & 1 & 0 & 0 & 0 & 3 \end{bmatrix}$$

$$\begin{bmatrix} 1 & 1 & 0 & 1 & 0 & 2 \\ 1 & 0 & 1 & 1 & 0 & 2 \\ 1 & 1 & 0 & 0 & 1 & 3 \\ 1 & 0 & 1 & 0 & 1 & 3 \\ 1 & 1 & 0 & 0 & 0 & 0 \\ 1 & 0 & 1 & 0 & 0 & 0 \end{bmatrix}$$

7. Consider the system

$$2x + 3y = 1$$
$$4x + 6y = 3$$

Notice that the right hand side is not in the column space of the coefficient matrix.

(a) Find the point in the column space that is closest to $\begin{bmatrix} 1 \\ 3 \end{bmatrix}$.

(b) Solve the system that arises when the point found in (a) is used as the right hand side.

(c) The answer to (b) is an entire line of solutions. Find the solution which is closest to the origin. (This can be done either geometrically or as a constrained extremal problem: you are to minimize distance from the origin — basically $x^2 + y^2$ — subject to (x,y) being on the line you found in (b).)

3.2. The Least Squares Solution of Overdetermined Linear Systems

Systems having more equations than unknowns often arise in data fitting problems. In the simplest case a researcher might assume that the output, y, of an experiment is a linear function of the input, x. In other words,

$$y = ax + b$$

is assumed where the parameters a and b are to be found to fit the observed experimental data. Typically, the experiment is performed many times. This generates data $(x_1,y_1),(x_2,y_2),\ldots,(x_N,y_N)$ which are supposed to satisfy

$$y_i = ax_i + b, \quad i = 1,2,3,\ldots,N, \tag{1}$$

for some numbers a and b. This can be viewed as a system of linear equations for a and b:

$$\begin{bmatrix} x_1 & 1 \\ x_2 & 1 \\ \vdots & \vdots \\ x_N & 1 \end{bmatrix} \begin{bmatrix} a \\ b \end{bmatrix} = \begin{bmatrix} y_1 \\ \vdots \\ \vdots \\ y_N \end{bmatrix} \tag{2}$$

If $N \geq 3$, then (2) will in general have no solution. For example, in the $N \geq 3$ case the column space of the matrix is either a line or a plane and the chance that the vector of outputs $\begin{bmatrix} y_1 \\ y_2 \\ y_3 \end{bmatrix}$ is in the column space is small. In higher dimensions it doesn't get better. Nevertheless, there is the need to make some good sense out of (2).

One standard way of treating systems like (2) is to replace the right hand side by a vector in the column space of the coefficient matrix. The best vector to choose would be the member of the column space closest to the right hand side. We formulate this as a general principle.

Least Squares Principle. If the system $A\mathsf{x} = \mathsf{b}$ has no solution, then replace b by the member of the column space closest to $\mathbf{b}$, say $\mathbf{b}^*$, and then solve $A\mathbf{x} = \mathbf{b}^*$. If $A\mathbf{x} = \mathbf{b}^*$ has a unique solution, it is called the *least squares solution* of $A\mathbf{x} = \mathbf{b}$.

There are still problems to be overcome. First, how do we find $\mathbf{b}^*$? Next, what should we do if $A\mathbf{x} = \mathbf{b}^*$ does not have a unique solution?

Our present treatment avoids the first problem and ignores the second. However, in some simple cases, like system (2), one can readily check if $A\mathbf{x} = \mathbf{b}^*$ has a unique solution.

Theorem 1. If A is an $N\times 2$ matrix and neither column is a multiple of the other, then the system $A\mathbf{x} = \mathbf{b}$ has either no solutions or exactly one solution.

The following result is more general but the hypotheses are not easy to check.

Theorem 2. If A is an $n\times k$ matrix and if no column is a linear combination of the others, then the system $A\mathbf{x} = \mathbf{b}$ has either no solutions or exactly one solution.

Remark. The condition on the columns in Theorem 2 is called *linear independence.* □

Under the conditions of Theorem 1 or Theorem 2, there will be a unique solution to $A\mathbf{x} = \mathbf{b}^*$.

Now, consider the case of a 3×2 matrix whose column space is a plane. From basic facts about projections onto planes, we have

Fact. $\mathbf{b}^*$ is the point in the column space of A closest to $\mathbf{b}$ if and only if $\mathbf{b}-\mathbf{b}^*$ is perpendicular to every vector in the column space of A. That is, for every $\mathbf{z}$ in $\mathbb{R}^2$, $A\mathbf{z}$ is perpendicular to $\mathbf{b}-\mathbf{b}^*$:

$$(A\mathbf{z})^t(\mathbf{b}-\mathbf{b}^*) = 0. \tag{3}$$

Since we know $\mathbf{b}^*$ is in the column space of A, we can write $A\mathbf{x}^* = \mathbf{b}^*$. Also, we have

$$(A\mathbf{z})^t = \mathbf{z}^t A^t$$

so (3) becomes

$$\mathbf{z}^t A^t(\mathbf{b} - A\mathbf{x}^*) = 0$$

for all $\mathbf{z} \in \mathbf{R}^2$. But this says that $A^t(\mathbf{b} - A\mathbf{x}^*)$ is a vector in $\mathbf{R}^2$ which is perpendicular to *every* vector in $\mathbf{R}^2$. The only way this can be is if

$$A^t\mathbf{b} - A^tA\mathbf{x}^* = 0. \tag{4}$$

Notice that A^t is a 2×3 matrix so A^tA is a 2×2 matrix. So, system (4) is a 2×2 system for $\mathbf{x}^*$:

$$A^tA\mathbf{x}^* = A^t\mathbf{b}. \tag{5}$$

The equations (5) are called the *normal equations.* Although the above development was phrased in terms of a 3×2 system, it is valid for general overdetermined systems.

Theorem 3. Let A be an $n \times k$ matrix whose columns are linearly independent (i.e., satisfy the hypotheses of Theorem 2), then the least squares solution of $A\mathbf{x} = \mathbf{b}$ can be found by solving the $k \times k$ system

$$(A^tA)\mathbf{x}^* = A^t\mathbf{b}.$$

Example 1. Find the least squares solution of

$$\begin{aligned} x + 2y &= 3 \\ y &= 4 \\ x + y &= 10. \end{aligned}$$

(See Example 4, Section 3.1.)

Solution.

$$A = \begin{bmatrix} 1 & 2 \\ 0 & 1 \\ 1 & 1 \end{bmatrix}, \qquad A^t = \begin{bmatrix} 1 & 0 & 1 \\ 2 & 1 & 1 \end{bmatrix}$$

so

$$A^tA = \begin{bmatrix} 2 & 3 \\ 3 & 6 \end{bmatrix} \quad \text{and} \quad A^t\mathbf{b} = \begin{bmatrix} 13 \\ 20 \end{bmatrix}.$$

The solution of

$$\begin{aligned} 2x + 3y &= 13 \\ 3x + 6y &= 20 \end{aligned}$$

is $x = 6$, $y = \frac{1}{3}$. □

Example 2. The data in the table below appear to lie near a straight line $y = mx + b$. Find the values of m and b that give the least squares fit to the data.

x_i	0.78	0.80	0.82	0.84	0.86	0.88
y_i	0.70	0.72	0.73	0.74	0.76	0.77

Solution. The linear system for m and b is given by the equations

$$mx_i + b = y_i \quad i = 1,2,\ldots,6.$$

$$\begin{bmatrix} 0.78 & 1 \\ 0.80 & 1 \\ 0.82 & 1 \\ 0.84 & 1 \\ 0.86 & 1 \\ 0.88 & 1 \end{bmatrix} \begin{bmatrix} m \\ b \end{bmatrix} = \begin{bmatrix} 0.70 \\ 0.72 \\ 0.73 \\ 0.74 \\ 0.76 \\ 0.77 \end{bmatrix}.$$

Now,

$$A^t = \begin{bmatrix} 0.78 & 0.80 & 0.82 & 0.84 & 0.86 & 0.88 \\ 1 & 1 & 1 & 1 & 1 & 1 \end{bmatrix}$$

$$A^t\mathbf{y} = \begin{bmatrix} 3.6734 \\ 4.4200 \end{bmatrix} \text{ and } A^tA = \begin{bmatrix} 4.1404 & 4.9800 \\ 4.9800 & 6 \end{bmatrix}.$$

So, we must solve

$$4.1404m^* + 4.98b^* = 3.6734$$
$$4.98m^* + 6b^* = 4.42.$$

With a hand calculator we obtain $m^* = 0.6857$ and $b^* = 0.1675$. Here is a sketch of the straight line together with the data points.

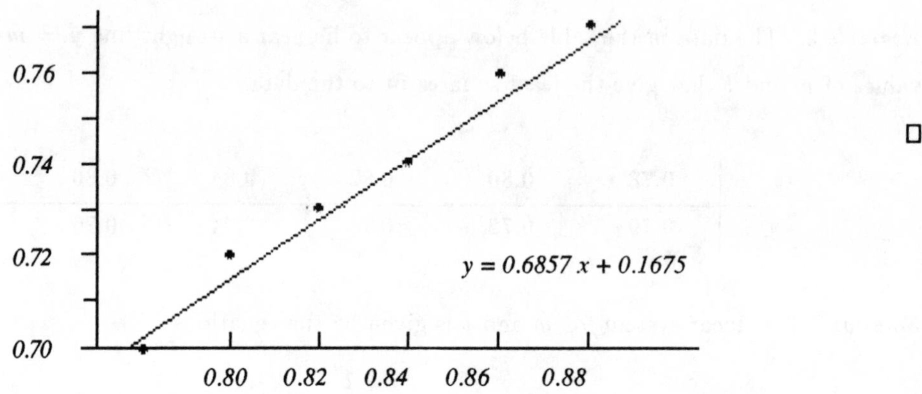

□

Example 3. A researcher obtained the data in the table below and suspects that the data points all lie on or near a quadratic so it is reasonable to set up the system

$$y_i = ax_i^2 + bx_i + c, \qquad i = 1,2,...,6$$

and find the least squares solution a^*, b^*,c^*.

i	x_i	y_i
1	0	5
2	2	11
3	4	32
4	6	73
5	8	122
6	10	200

Solution. We have a 6×3 system, $A\mathbf{x} = \mathbf{y}$, where the entries of $\mathbf{y}$ can be read directly from the table, the last column of A contains all 1's, the middle column contains the x_i's and the first column contains the squares of the x_i's.

$$\begin{bmatrix} 0 & 0 & 1 \\ 4 & 2 & 1 \\ 16 & 4 & 1 \\ 36 & 6 & 1 \\ 64 & 8 & 1 \\ 100 & 10 & 1 \end{bmatrix} \begin{bmatrix} a \\ b \\ c \end{bmatrix} = \begin{bmatrix} 5 \\ 11 \\ 32 \\ 73 \\ 122 \\ 200 \end{bmatrix}.$$

We compute

$$A^tA = \begin{bmatrix} 15{,}664 & 1{,}800 & 220 \\ 1{,}800 & 220 & 30 \\ 220 & 30 & 6 \end{bmatrix} \quad \text{and} \quad A^t\mathbf{y} = \begin{bmatrix} 30{,}992 \\ 3{,}564 \\ 443 \end{bmatrix}$$

and solve

$$\begin{aligned} 15{,}664a^* + 1800b^* + 220c^* &= 30{,}992 \\ 1{,}800a^* + 220b^* + 30c^* &= 3{,}564 \\ 220a^* + 30b^* + 6c^* &= 443. \end{aligned}$$

From a programmable hand calculator we get $a^* = 2.11$, $b^* = -1.80$, $c^* = 5.57$.

3.2. EXERCISES

1. Find the least squares solution of $A\mathbf{x} = \mathbf{b}$ if

(a) $A = \begin{bmatrix} 2 & 1 \\ 4 & 2 \\ 1 & 1 \end{bmatrix}$, $\quad \mathbf{b} = \begin{bmatrix} 1 \\ 0 \\ 0 \end{bmatrix}$, $\quad \mathbf{b} = \begin{bmatrix} 0 \\ 1 \\ 0 \end{bmatrix}$

(b) $A = \begin{bmatrix} 1 & 3 \\ -1 & 1 \\ 2 & 1 \end{bmatrix}$, $\quad \mathbf{b} = \begin{bmatrix} 0 \\ 1 \\ 1 \end{bmatrix}$, $\quad \mathbf{b} = \begin{bmatrix} 1 \\ 2 \\ 3 \end{bmatrix}$

2. For each system in Exercise 1, verify that $\mathbf{b} - A\mathbf{x}^*$ is perpendicular to the columns of A.

3. Find the straight line $y = mx + b$ that fits the data below in the least squares sense.

x_i	0	1	2	3	4	5	6
y_i	-3	0	3	5	7	12	14

4. The data table in Example 2 was generated by taking $y_i = \sin x_i$ for the given values of x_i (in radians). Check how well $\sin x$ is approximated by $0.6857x + 0.1675$ in the range $.68 \leq x \leq .94$ by computing the error at various points in this range.

5. Use the known values of $\cos x$ for $x = 0, \frac{\pi}{6}, \frac{\pi}{4}, \frac{\pi}{2}$ to find a quadratic that fits this data in the least squares sense. Compare the true value of $\cos x$ with the values of the quadratic at $x = 0, \frac{\pi}{12}, \frac{2\pi}{12}, \ldots, \frac{6\pi}{12}$.

6. Find the least squares solution of the system

$$\begin{aligned} x + 2y \phantom{{}+z} &= 0 \\ 2x + y + z &= 1 \\ 2y + z &= 3 \\ x + y + z &= 0 \\ 3x \phantom{{}+y} + 2z &= -1. \end{aligned}$$

7. The columns of matrix $B = \begin{bmatrix} 1 & 2 \\ 2 & 4 \\ 3 & 6 \end{bmatrix}$ are not linearly independent. Show that the matrix B^tB has a line as its range.

8. The columns of $A = \begin{bmatrix} 1 & 1 \\ 0.00001 & 0 \\ 0 & 0.00001 \end{bmatrix}$ are linearly independent. Compute A^tA and round off the entries to six decimal places. Let $\tilde{A}$ denote the resulting matrix. Does the system $\tilde{A}\mathbf{x} = \mathbf{b}$ always have a solution? Does the unrounded system $A^tA\mathbf{x} = \mathbf{b}$ always have a solution?

9. Let A be as in Theorem 1 and define $A^\dagger = (A^tA)^{-1}A^t$. $A^\dagger$ is called the *Generalized Inverse of* A. Verify that $A^\dagger A = I$ and $(AA^\dagger)(AA^\dagger) = AA^\dagger$. Is $AA^\dagger = I$?

3.3. The Null Space of a Matrix

We now study the geometry of the solution set of $A\mathsf{x} = \mathsf{b}$. If the system has exactly two unknowns and A is not the zero matrix, then the solution set is either empty, a single point, or a line. For nontrivial systems with three unknowns the solution set is either empty, a point, a line, or a plane. The geometric structure can be related to the solution set of the "homogeneous" system $A\mathsf{x} = \mathbf{0}$. We will use our knowledge of the geometric nature of the solution set to solve the problem of finding the "best" solution when there are infinitely many solutions.

Example 1. Let $A = \begin{bmatrix} 1 & 2 \\ 2 & 4 \end{bmatrix}$ and assume b is in the column space of A. Find the solution set of $A\mathsf{x} = \mathsf{b}$; interpret the solution set geometrically, and find the "best" solution for such b.

Solution. First we note that the column space of A is the line through the origin with slope 2; so we consider b with $b_2 = 2b_1$. Before looking at the general case, we'll look at some specific b's.

Let $b_1 = b_2 = 0$. Then, $A\mathsf{x} = \mathsf{b}$ is simply

$$\begin{aligned} x + 2y &= 0 \\ 2x + 4y &= 0. \end{aligned}$$

Since the equations are equivalent, we see that the solution set is the line $x + 2y = 0$. In other words, A takes the line $x + 2y = 0$ into the point (0,0). That is, *every* point p on this line satisfies $A\mathsf{p} = 0$.

We next consider $b_1 = 3$, $b_2 = 6$. Writing $\mathsf{b} = A\mathsf{x}$, we see

$$\begin{bmatrix} 3 \\ 6 \end{bmatrix} = \begin{bmatrix} 1 & 2 \\ 2 & 4 \end{bmatrix}\begin{bmatrix} x \\ y \end{bmatrix} = x\begin{bmatrix} 1 \\ 2 \end{bmatrix} + y\begin{bmatrix} 2 \\ 4 \end{bmatrix} = (x + 2y)\begin{bmatrix} 1 \\ 2 \end{bmatrix}.$$

Therefore, any x and y for which $x + 2y = 3$ gives a solution. A takes the line $x + 2y = 3$ in the $x-y$ plane into the single point (3,6) in the $b_1 - b_2$ plane. See Figure 1.

For general b in the column space of A we have $b_1 = p$ and $b_2 = 2p$. From $\mathsf{b} = A\mathsf{x}$ we see

$$p\begin{bmatrix} 3 \\ 6 \end{bmatrix} = \begin{bmatrix} 1 & 2 \\ 2 & 4 \end{bmatrix}\begin{bmatrix} x \\ y \end{bmatrix} = (x + 2y)\begin{bmatrix} 1 \\ 2 \end{bmatrix}.$$

So, A takes the line $x + 2y = p$ to the point $(p,2p)$. See Figure 2.

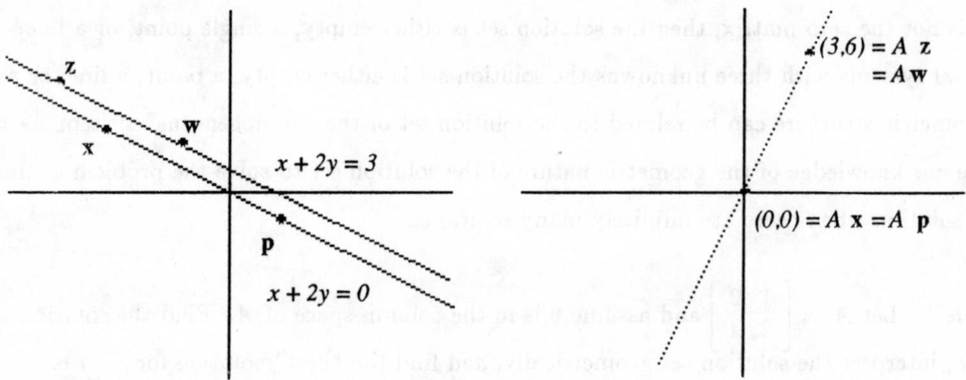

Figure 1

A takes every point on $x + 2y = 0$ to $\begin{bmatrix} 0 \\ 0 \end{bmatrix}$ and every point on $x + 2y = 3$ to $\begin{bmatrix} 3 \\ 6 \end{bmatrix}$.

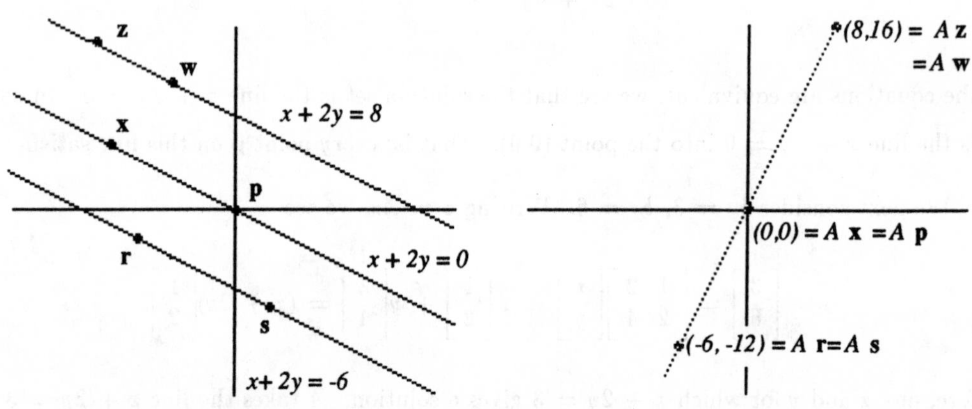

Figure 2. Every point on $x + 2y = p$ goes to $\begin{bmatrix} p \\ 2p \end{bmatrix}$.

Finally, the problem of finding the best solution is solved as follows. Take the point in the solution set which is closest to the origin. If the solution set is $x + 2y = p$, then we want the point on this line with the property that the line connecting it to the origin is perpendicular to $x + 2y = 0$.

Since vectors perpendicular to $x + 2y = 0$ are multiples of the normal vector (1,2), we see that the best solution must satisfy

$$x = t$$
$$y = 2t$$
$$x + 2y = p$$

So, $t + 4t = p$ or $t = \frac{1}{5}p$. Therefore, in general

$$x = \tfrac{1}{5}p \text{ and } y = \tfrac{2}{5}p.$$

For the system

$$x + 2y = 3$$
$$2x + 4y = 6$$

the best solution is $x = \frac{3}{5}$, $y = \frac{6}{5}$. □

There are a several important ideas contained in Example 1. The most important is related to the fact that the solution sets for different right hand sides are all parallel to the solution of $A\mathbf{x} = \mathbf{0}$. This set plays an important role in the general theory.

Definition. Let A be an $n{\times}k$ matrix. The set of all solutions of the system $A\mathbf{x} = \mathbf{0}$ is called the *null space* of A.

The null space of A is sometimes called the *kernel* of A. Notice that the zero vector is always in the null space. A system in which the right hand side is the zero vector is called a *homogeneous* system.

The fact that the solution sets in Example 1 were all parallel to the solution set of the homogeneous system is representative of the general case. Consider first the following examples.

Example 2. Let $A = \begin{bmatrix} 1 & 2 & 3 \\ 2 & 4 & 6 \\ 3 & 6 & 9 \end{bmatrix}$. Find the null space of A and show that the solution sets of $A\mathbf{x} = \mathbf{b}$

are planes parallel to the null space.

Solution. To find the null space we solve $A\mathbf{x} = \mathbf{0}$:

$$\begin{aligned} x + 2y + 3z &= 0 \\ 2x + 4y + 6z &= 0 \\ 3x + 6y + 9z &= 0. \end{aligned}$$

The second equation is twice the first and the third is three times the first so the system is equivalent to the single equation

$$x + 2y + 3z = 0.$$

So, (x,y,z) is in the null space if and only if it lies on the plane $x + 2y + 3z = 0$.

Now, if $\mathbf{b}$ is not in the column space of A, there is no solution to $A\mathbf{x} = \mathbf{b}$. So we consider only $\mathbf{b}$'s in the column space. Our observations concerning how the equations were multiples of each other show that the column space is the line $t\begin{bmatrix} 1 \\ 2 \\ 3 \end{bmatrix}$. Now, if $b_1 = t$, $b_2 = 2t$, and $b_3 = 3t$, we have

$$\begin{aligned} x + 2y + 3z &= t \\ 2x + 4y + 6z &= 2t \\ 3x + 6y + 9z &= 3t. \end{aligned}$$

These equations are all equivalent to the equation

$$x + 2y + 3z = t$$

which is a plane parallel to the null space $x + 2y + 3z = 0$. Thus, the solution sets are all planes parallel to the null space. □

***Example* 3.** Let $A = \begin{bmatrix} 1 & 4 & 7 \\ 2 & 5 & 8 \\ 3 & 6 & 9 \end{bmatrix}$. Find the null space of A and show that the solution sets of $A\mathbf{x} = \mathbf{b}$ are lines parallel to the null space.

Solution. First we solve $A\mathbf{x} = \mathbf{0}$:

$$\begin{aligned} x + 4y + 7z &= 0 \\ 2x + 5y + 8z &= 0 \\ 3x + 6y + 9z &= 0. \end{aligned}$$

Eliminating x from the second and third equations, we get

$$\begin{aligned} x + 4y + 7z &= 0 \\ -3y - 6z &= 0 \\ -6y - 12z &= 0. \end{aligned}$$

The second and third equations are equivalent to $y = -2z$. So, setting $z = t$, we get $y = -2t$ and $x = -7t - 4(-2t) = t$. So (x,y,z) is in the null space if and only if there is a number t so that $x = t$, $y = -2t$ and $z = t$. Geometrically, the null space is the line

$$\begin{bmatrix} x \\ y \\ z \end{bmatrix} = t \begin{bmatrix} 1 \\ -2 \\ 1 \end{bmatrix}$$

through the origin in the direction of $(1,-2,1)$.

The column space of A is the plane $x - 2y + z = 0$. {This is because the middle row of the matrix is the average of the first and the last.} If we solve the system

$$\begin{aligned} x + 4y + 7z &= a \\ 2x + 5y + 8z &= b \\ 3x + 6y + 9z &= c \end{aligned}$$

and assume $a - 2b + c = 0$, we get, by Gaussian Elimination,

$$\begin{aligned} x + 4y + 7z &= a \\ -3y - 6z &= b - 2a \\ -6y - 12z &= c - 3a. \end{aligned}$$

There will be a solution to the last two equations only if $2(b-2a) = c - 3a$. But, this just requires $2b = c + a$ which is what we've assumed. So, we can set $z = t$ and obtain

$$z = t$$
$$y = \frac{2a - b - 6t}{3} = -2t + \tfrac{2}{3}a - \tfrac{1}{3}b.$$
$$x = t - \tfrac{5}{3}a + \tfrac{4}{3}b.$$

So, the solution set is the line

$$\begin{bmatrix} x \\ y \\ z \end{bmatrix} = t\begin{bmatrix} 1 \\ -2 \\ 1 \end{bmatrix} + \begin{bmatrix} -\frac{5}{3}a + \frac{4}{3}b \\ \frac{2}{3}a - \frac{1}{3}b \\ 0 \end{bmatrix}$$

which is parallel to the null space. □

In each of the above examples the solution set was parallel to the null space of the coefficient matrix. In fact, once one solution was known all other solutions could have been found simply by adding on the general members of the null space. This is the content of

Theorem 1. If the $n \times k$ system $A\mathbf{x} = \mathbf{b}$ has a solution $\mathbf{x}_0$, then the complete solution set is

$$\{\mathbf{x}_0 + \mathbf{z}\text{: } \mathbf{z} \textit{ is in the null space of } A\}.$$

Corollary. If the null space of A consists of only the zero vector and if $A\mathbf{x} = \mathbf{b}$ has a solution, then the solution is unique.

Example 4. The null space of the matrix $\begin{bmatrix} 1 & 1 & 1 \\ 1 & 2 & 1 \end{bmatrix}$ is the line $\begin{bmatrix} x \\ y \\ z \end{bmatrix} = t\begin{bmatrix} 1 \\ 0 \\ -1 \end{bmatrix}$. Find the complete solution set of the system

$$x + y + z = 2$$
$$x + 2y + z = 2$$

if one solution is $x = 2,\ y = 0,\ z = 0$.

Solution. Theorem 1 says that every member of the solution set is of the form

$$\begin{bmatrix} x \\ y \\ z \end{bmatrix} = \begin{bmatrix} 2 \\ 0 \\ 0 \end{bmatrix} + t\begin{bmatrix} 1 \\ 0 \\ -1 \end{bmatrix}.$$ □

Example 5. Show that if A is a $k\times 2$ matrix whose columns are not multiples of each other, then $A\mathbf{x} = \mathbf{b}$ has either exactly one solution or no solution.

Solution. The system has no solution if $\mathbf{b}$ is not in the column space of A. To show that there is exactly one solution otherwise, we need only show that the null space of A contains only the zero vector. Say $A\mathbf{x} = \mathbf{0}$. Let the first equation of this system be

$$ax + by = 0.$$

Now, there is another equation in the system, say

$$cx + dy = 0$$

with $\frac{a}{b} \neq \frac{c}{d}$, or $ad - bc \neq 0$. From the basic results on inverses of 2×2 matrices we see that $\begin{bmatrix} a & b \\ c & d \end{bmatrix}$ has an inverse so that

$$\begin{bmatrix} x \\ y \end{bmatrix} = \begin{bmatrix} a & b \\ c & d \end{bmatrix}^{-1}\begin{bmatrix} 0 \\ 0 \end{bmatrix} = \begin{bmatrix} 0 \\ 0 \end{bmatrix}.$$ □

Like the column space of A, the null space has dimension. The important connection between the dimensions of the column and null spaces is

The Law of Conservation of Dimension. If A is a matrix having k-columns, then

the number of columns = (*dimension of column space*) + (*dimension of null space*).

In Example 1, we had a matrix with two columns. The null space and column space were each lines.

In Example 2, we had a matrix with three columns. The column space was a line so the null space *had to be a plane.*

In Example 3, the matrix had three columns. The null space was a line and the column space was a plane.

If we have a square system, n equations in n unknowns, then the column space is a subset of $\mathbb{R}^n$ and the only way the column space can fail to be all of $\mathbb{R}^n$ is if the dimension of the null space is nonzero. This is the essence of

Theorem 2. Let A be an $n \times n$ matrix. The system $A\mathbf{x} = \mathbf{b}$ has a solution for every $\mathbf{b}$ in $\mathbb{R}^n$ if and only if the only solution of $A\mathbf{x} = \mathbf{0}$ is $\mathbf{x} = \mathbf{0}$.

Example 6. Explain why the system below must have either no solutions or infinitely many solutions.

$$\begin{aligned} 2x + 3y + z + w &= 4 \\ x - y + z + w &= 82 \\ -x - y - z + 2w &= 10 \end{aligned}$$

Solution. The coefficient matrix has four columns. But the column space is a subset of $\mathbb{R}^3$ so its dimension must be at most three. Now,

$$\begin{aligned} \textit{dimension of null space} &= (\#\ \textit{of columns}) - (\textit{dimension of column space}) \\ &= 4 - (\textit{dimension of column space}) \\ &\geq 4 - 3 = 1. \end{aligned}$$

If there is a solution, then there is at least a line of solutions. In fact the column space has dimension 3 (this is an exercise for the reader) so the solution set is a line. □

3.3. EXERCISES

1. Find the range and null space of each of the following 2×2 matrices.

(a) $\begin{bmatrix} 2 & 3 \\ 1 & 4 \end{bmatrix}$ (b) $\begin{bmatrix} 0 & 1 \\ 1 & 0 \end{bmatrix}$ (c) $\begin{bmatrix} 4 & 5 \\ 8 & 10 \end{bmatrix}$ (d) $\begin{bmatrix} -1 & 2 \\ 3 & -6 \end{bmatrix}$

(e) $\begin{bmatrix} 2 & 2 \\ -1 & -1 \end{bmatrix}$ (f) $\begin{bmatrix} 0 & 1 \\ 0 & 0 \end{bmatrix}$ (g) $\begin{bmatrix} 1 & 0 \\ 0 & 0 \end{bmatrix}$

2. For each part of Problem 1 for which the range was not all of $\mathbb{R}^2$, sketch the range and null space. In which examples are the range and null space mutually perpendicular?

3. Find the range and null space of each of the following and verify the Law of Conservation of Dimension in each case.

(a) $\begin{bmatrix} 1 & 2 & 3 & 4 \\ 5 & 6 & 7 & 8 \\ 9 & 10 & 11 & 12 \end{bmatrix}$ (b) $\begin{bmatrix} 1 & 0 & 0 \\ 2 & 2 & 3 \\ 0 & 1 & -1 \end{bmatrix}$ (c) $\begin{bmatrix} 2 & 2 \\ 1 & 1 \\ 3 & 3 \end{bmatrix}$

4. If a system of linear equations with more unknowns than equations has a solution, then it must have infinitely many solutions. Explain how this is a consequence of the law of conservation of dimension.

5. In Example 5 it was implicitly assumed that neither b nor d was zero. Show that the statement of Example 5 is true even if b or d is zero.

6. Can the column space of a 2×2 matrix be the same as its null space? Look at the results of Problem 1 for guidance.

7. Can the column space of a 3×3 matrix be the same as its null space?

8. Find a 3×3 matrix whose column space is a subset of its null space.

3.4. Linear Independence

The term "linear independence" was introduced in (the optional) Section 3.2. We will discuss this fundamental concept and its relation to null spaces and linear systems.

Definition. The vectors $\mathbf{v}_1,\mathbf{v}_2,\ldots,\mathbf{v}_p$ are *linearly dependent* if one of them can be expressed as a linear combination of the others.

Example 1. The vectors (1,1), (−1,1), (0,4) are linearly dependent since

$$\begin{bmatrix} 0 \\ 4 \end{bmatrix} = 2\begin{bmatrix} 1 \\ 1 \end{bmatrix} + 2\begin{bmatrix} -1 \\ 1 \end{bmatrix}; \tag{1}$$

equivalently,

$$2\begin{bmatrix} 1 \\ 1 \end{bmatrix} + 2\begin{bmatrix} -1 \\ 1 \end{bmatrix} - \begin{bmatrix} 0 \\ 4 \end{bmatrix} = \begin{bmatrix} 0 \\ 0 \end{bmatrix}. \tag{2}$$

□

Definition. The vectors $\mathbf{v}_1,\mathbf{v}_2,\ldots,\mathbf{v}_p$ are *linearly independent* if they are not linearly dependent.

Intuitively, a set of vectors is linearly independent if it contains no redundancies with respect to the fundamental operation of linear algebra — that of forming linear combinations.

Example 2. Two vectors $\mathbf{v}$ and $\mathbf{u}$ are linearly independent if they are not multiplies of each other. This is because one is a linear combination of the other if and only if it is a multiple of the other.

In many applications, the linear dependence or independence of a set of vectors is a known consequence of the science and mathematics inherent in the application. For example in $\mathbb{R}^3$ any three mutually perpendicular vectors must be linearly independent — any linear combination of two of them is constrained to lie in the plane perpendicular to the third. More generally, we have

Theorem 1. Let $\mathbf{v}_1,\mathbf{v}_2,\ldots,\mathbf{v}_p$ be non-zero vectors in $\mathbb{R}^n$ with $\mathbf{v}_i^t\mathbf{v}_j = 0$ for $i \neq j$, then these vectors are linearly independent and $p \leq n$.

Example 2. In $\mathbb{R}^4$ consider (1,1,1,1), (2,−2,0,0), (0,0,1,−1), (1,1,−1,−1). All pairwise inner products are 0, so these vectors are linearly independent.

In overdetermined linear systems one often knows that the coefficient matrix has linearly independent columns. The fact that the normal equations $A^tA\mathbf{x} = A^t\mathbf{b}$ have a unique solution is a consequence of

Theorem 2. If the columns of the $n \times k$ matrix A are linearly independent, then the columns of the $k \times k$ matrix A^tA are also linearly independent.

A standard way of testing for linear independence is a consequence of

Theorem 3. The vectors $\mathbf{v}_1, \mathbf{v}_2, \ldots, \mathbf{v}_p$ are linearly independent if and only if the only linear combination of the $\mathbf{v}_i$'s yielding the zero vector is the linear combination $0\mathbf{v}_1 + 0\mathbf{v}_2 + \cdots + 0\mathbf{v}_p$. That is $\mathbf{v}_1, \ldots, \mathbf{v}_p$ are linearly independent if and only if

$$\alpha_1\mathbf{v}_1 + \alpha_2\mathbf{v}_2 + \cdots + \alpha_p\mathbf{v}_p = \mathbf{0}$$

forces

$$\alpha_1 = \alpha_2 = \cdots = \alpha_p = 0.$$

The verification of this important fact uses an abstraction of the algebra that led from (1) to (2) in Example 1 together with the definition.

Example 3. Use Theorem 2 to show that the vectors of Example 2 are linearly independent.

Solution. Assume that

$$\alpha_1\begin{bmatrix} 1 \\ 1 \\ 1 \\ 1 \end{bmatrix} + \alpha_2\begin{bmatrix} 2 \\ -2 \\ 0 \\ 0 \end{bmatrix} + \alpha_3\begin{bmatrix} 0 \\ 0 \\ 1 \\ -1 \end{bmatrix} + \alpha_4\begin{bmatrix} 1 \\ 1 \\ -1 \\ -1 \end{bmatrix} = \begin{bmatrix} 0 \\ 0 \\ 0 \\ 0 \end{bmatrix}.$$

Observe that this vector equation is equivalent to the homogeneous linear system whose coefficient matrix has the given vectors as columns

$$\begin{bmatrix} 1 & 2 & 0 & 1 \\ 1 & -2 & 0 & 1 \\ 1 & 0 & 1 & -1 \\ 1 & 0 & -1 & -1 \end{bmatrix}\begin{bmatrix} \alpha_1 \\ \alpha_2 \\ \alpha_3 \\ \alpha_4 \end{bmatrix} = \begin{bmatrix} 0 \\ 0 \\ 0 \\ 0 \end{bmatrix}. \tag{3}$$

One can now check with Gaussian Elimination that this system has only the solution $\alpha_1 = \alpha_2 = \alpha_3 = \alpha_4 = 0$ (see Exercise 1). □

This example demonstrates a relation between the linear independence of a set of vectors and the null space of an associated matrix and thus suggests a method for testing linear independence —

compute the null space.

Theorem 4. The vectors $\mathbf{v}_1,\mathbf{v}_2,\ldots,\mathbf{v}_p$ are linearly independent if and only if the null space of the matrix $[\mathbf{v}_1 \;\; \mathbf{v}_2 \;\; \cdots \;\; \mathbf{v}_p]$ is $\{\mathbf{0}\}$.

One can now *rephrase* some of the results of the preceding sections in terms of linear independence. For example Theorem 2 of Section 3.3 is equivalent to

Theorem 5. Let A be an $n \times n$ matrix. The system $A\mathbf{x} = \mathbf{b}$ has a unique solution for every $\mathbf{b}$ in $\mathbb{R}^n$ if and only if the columns of A are linearly independent.

3.4. EXERCISES

1. Show that the only solution to system (3) in Example 3 is the zero solution.

2. Are the columns of $\begin{bmatrix} 1 & 4 & 7 \\ 2 & 5 & 8 \\ 3 & 6 & 9 \end{bmatrix}$ linearly independent?

3. If $\mathbf{v}_1,\mathbf{v}_2,\mathbf{v}_3$ are in $\mathbb{R}^2$, then they must be linearly dependent. Explain this geometrically and algebraically.

4. If $\mathbf{v}_1,\mathbf{v}_2,\ldots,\mathbf{v}_p$ are in $\mathbb{R}^n$ and $p > n$, then they must be linearly dependent. Explain how this is a consequence of the Law of Conservation of Dimension applied to the matrix $[\mathbf{v}_1 \;\; \mathbf{v}_2 \;\; \cdots \;\; \mathbf{v}_p]$.

5. Below are the steps that prove Theorem 2. Give a reason for each step. Suppose $\mathbf{x}$ is in the null space of A^tA, then

   (a) $A^tA\mathbf{x} = \mathbf{0}$ (b) $\mathbf{x}^tA^tA\mathbf{x} = 0$

   (c) $(A\mathbf{x})^tA\mathbf{x} = 0$ (d) $|A\mathbf{x}| = 0$

   (e) $\mathbf{x}$ is in the null space of A (f) $\mathbf{x} = \mathbf{0}$

   (g) the columns of A^tA are linearly independent.

3.5. The Least Squares Solution of Underdetermined Linear Systems

We consider the system $A\mathbf{x} = \mathbf{b}$ where A is an $n \times k$ matrix with $n < k$ and develop a method for finding the unique solution closest to the origin in the case that the column space of A is $\mathbb{R}^n$. The null space plays an important role.

If A is a 2×3 matrix whose rows are not multiples of each other, then the solution set of the system $A\mathbf{x} = \mathbf{b}$, for given $\mathbf{b}$ in $\mathbb{R}^2$, is a line parallel to the null space of A. See Example 5 of Section 3.3, for example. From Figure 1 one can see that the member of the solution set closest to the origin determines a vector *perpendicular* to the null space of A.

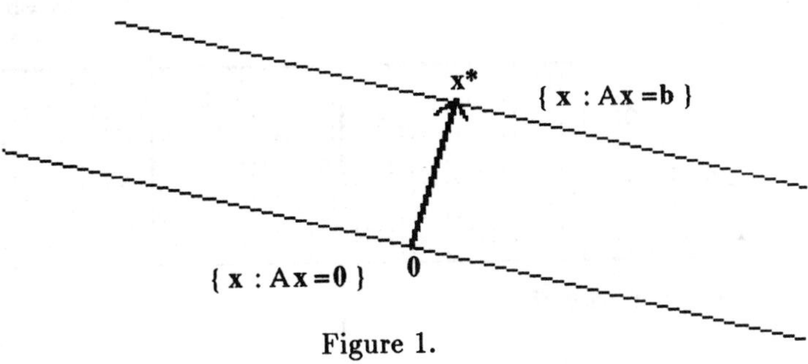

Figure 1.

Now, the null space is a line through the origin. The set of vectors perpendicular to this line form a plane. This plane must contain the columns of A^t (i.e., the rows of A) since members of the null space are perpendicular to the rows of A. But, the columns of A^t are not multiples of each other – this was an assumption. Therefore, the plane perpendicular to the null space of A is the plane determined by the columns of A^t. For matrices of arbitrary size we have

***Theorem* 1.** A vector $\mathbf{x}$ is perpendicular to the null space of A if and only if $\mathbf{x}$ is a member of the column space of A^t.

Since a member of the column space of A^t has the form $\mathbf{x} = A^t\mathbf{y}$, we can formulate a plan for computing $\mathbf{x}^*$, the member of the solution set closest to the origin.

Least Squares Principle for Underdetermined Systems. If A is an $n \times k$ matrix with $n < k$ and having linearly independent *row* vectors, then the unique solution of $A\mathbf{x} = \mathbf{b}$ which is closest to $\mathbf{0}$ satisfies

$$\mathbf{x}^* = A^t\mathbf{y} \tag{1}$$

where

$$AA^t\mathbf{y} = \mathbf{b}. \tag{2}$$

$\mathbf{x}^*$ is called the *least squares solution* of $A\mathbf{x} = \mathbf{b}$.

Property (1) is a consequence of the discussion above and property (2) is simply the statement that $A\mathbf{x}^* = \mathbf{b}$ since $\mathbf{x}^* = A^t\mathbf{y}$. Figure 2 depicts the situation for a 2×3 matrix

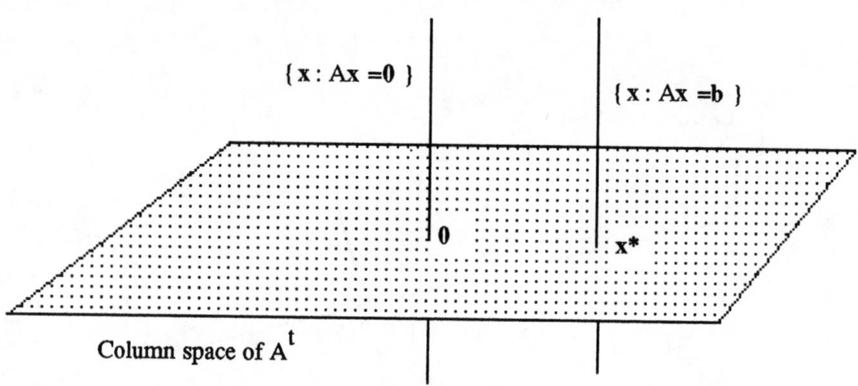

Figure 2. The point where the solution set of $A\mathbf{x} = \mathbf{b}$ intersects the column space of A^t is $\mathbf{x}^*$.

Example 1. Find the solution of

$$\begin{aligned} x_1 + x_2 + x_3 &= 1 \\ 2x_1 + x_2 \quad\;\; &= 1 \end{aligned}$$

that is closest to the origin.

Solution. The null space of $A = \begin{bmatrix} 1 & 1 & 1 \\ 2 & 1 & 0 \end{bmatrix}$ is clearly a line so the column space is $\mathbb{R}^2$. This means that the system has a line of solutions. First, we'll follow (2) and solve $AA^t\mathbf{y} = \mathbf{b}$; then, our solution will be $\mathbf{x}^* = A^t\mathbf{y}$.

Now,

$$AA^t = \begin{bmatrix} 1 & 1 & 1 \\ 2 & 1 & 0 \end{bmatrix}\begin{bmatrix} 1 & 2 \\ 1 & 1 \\ 1 & 0 \end{bmatrix} = \begin{bmatrix} 3 & 3 \\ 3 & 5 \end{bmatrix} \text{ and } \mathbf{b} = \begin{bmatrix} 1 \\ 1 \end{bmatrix}.$$

We solve

$$3y_1 + 3y_2 = 1$$
$$3y_1 + 5y_2 = 1$$

and get $y_1 = \frac{1}{3}$, $y_2 = 0$. Therefore,

$$\begin{bmatrix} x_1^* \\ x_2^* \\ x_3^* \end{bmatrix} = A^t\mathbf{y} = \begin{bmatrix} 1 & 2 \\ 1 & 1 \\ 1 & 0 \end{bmatrix} \begin{bmatrix} \frac{1}{3} \\ 0 \end{bmatrix} = \begin{bmatrix} \frac{1}{3} \\ \frac{1}{3} \\ \frac{1}{3} \end{bmatrix}$$ □

Remark. The method previously presented in Section 3.2 will handle problems of the form

$$A\mathbf{x} = \mathbf{b}$$

when A is $n \times k$ with $n > k$ and with linearly independent *columns*. The column space of A provides the geometry that motivates the ultimate algebraic computations. The method presented here takes care of the case $n < k$ if the *rows* are linearly independent. The null space provides the geometry.

More general least squares problems can be solved geometrically using a combination of these ideas. However, a development of the algebraic aspects of the general problem, the Singular Value Decomposition, is beyond the scope of this book.

***Example* 2.** Find the function of the form $f(x) = a + bx + ce^x$ that satisfies $f(0) = 0$, $f(1) = 1$ and minimizes $a^2 + b^2 + c^2$.

Solution. The given conditions require

$$0 = f(0) = a \qquad + c$$
$$1 = f(1) = a + b + ce.$$

There are infinitely many a, b, c that satisfy the equations. We are to find the solution closest to the origin in $\mathbb{R}^3$. Accordingly, we have

$$AA^t = \begin{bmatrix} 1 & 0 & 1 \\ 1 & 1 & e \end{bmatrix} \begin{bmatrix} 1 & 1 \\ 0 & 1 \\ 1 & e \end{bmatrix} = \begin{bmatrix} 2 & 1+e \\ 1+e & 2+e^2 \end{bmatrix}.$$

We solve

$$2y_1 + 3.7183y_2 = 0$$
$$3.7183\,y_1 + 9.3891y_2 = 1$$

and get $y_1 = -0.7508$, $y_2 = 0.4038$. Finally,

$$\begin{bmatrix} a \\ b \\ c \end{bmatrix} = \begin{bmatrix} 1 & 1 \\ 0 & 1 \\ 1 & e \end{bmatrix} \begin{bmatrix} y_1 \\ y_2 \end{bmatrix} = \begin{bmatrix} -0.3470 \\ 0.4038 \\ 0.3470 \end{bmatrix}.$$

So, the desired function is

$$f(x) = -0.3740 + 0.4038x + 0.3470e^x.$$ □

Example 3. Find the solution of

$$2x_1 - x_2 + x_3 = 3$$
$$x_1 + x_2 + x_3 + x_4 = 0$$

which is closest to the origin in $\mathbb{R}^4$.

Solution. First we notice that the column space of the coefficient matrix is $\mathbb{R}^2$ so the null space and the solution set must be two-dimensional subsets of $\mathbb{R}^4$. We *cannot* draw the intersection of the solution set with the column space of A^t; however, the Least Squares Principle for Underdetermined Systems still holds and we compute

$$AA^t = \begin{bmatrix} 2 & -1 & 1 & 0 \\ 1 & 1 & 1 & 1 \end{bmatrix} \begin{bmatrix} 2 & 1 \\ -1 & 1 \\ 1 & 1 \\ 0 & 1 \end{bmatrix} = \begin{bmatrix} 6 & 2 \\ 2 & 4 \end{bmatrix}$$

and solve

$$6y_1 + 2y_2 = 3$$
$$2y_1 + 4y_2 = 0$$

to get $y_1 = 0.6$ and $y_2 = -0.3$.

Hence, the desired solution vector is

$$\begin{bmatrix} x_1^* \\ x_2^* \\ x_3^* \\ x_4^* \end{bmatrix} = A^t\mathbf{y} = \begin{bmatrix} 2 & 1 \\ -1 & 1 \\ 1 & 1 \\ 0 & 1 \end{bmatrix} \begin{bmatrix} 0.6 \\ -0.3 \end{bmatrix} = \begin{bmatrix} 0.9 \\ -0.9 \\ 0.3 \\ -0.3 \end{bmatrix}. \qquad \square$$

The next example illustrates two points: first, the Least Squares solution can be viewed as the solution of a constrained extremal problem and, second, the method presented in this section applies to systems with more than two equations.

Example 4. Find the smallest value of $f(x_1,x_2,x_3,x_4) = x_1^2 + x_2^2 + x_3^2 + x_4^2$ subject to the equalities

$$\begin{aligned} x_1 + x_2 + x_3 + x_4 &= 0 \\ 2x_1 + x_2 \qquad\qquad &= 1 \\ x_2 + x_3 + 2x_4 &= 3. \end{aligned}$$

Solution. First we note that f is the square of the distance from the origin so we can treat this as the previous problems. The 3×4 coefficient matrix has $\mathbb{R}^3$ for its column space. So, the solution set is a one dimensional subset of $\mathbb{R}^4$. As before, it is impossible to draw a picture. But, the algebraic computations are still valid. So, we find

$$AA^t = \begin{bmatrix} 1 & 1 & 1 & 1 \\ 2 & 1 & 0 & 0 \\ 0 & 1 & 1 & 2 \end{bmatrix} \begin{bmatrix} 1 & 2 & 0 \\ 1 & 1 & 1 \\ 1 & 0 & 1 \\ 1 & 0 & 2 \end{bmatrix} = \begin{bmatrix} 4 & 3 & 4 \\ 3 & 5 & 1 \\ 4 & 1 & 6 \end{bmatrix}$$

and solve

$$\begin{aligned} 4y_1 + 3y_2 + 4y_3 &= 0 \\ 3y_1 + 5y_2 + y_3 &= 1 \\ 4y_1 + y_2 + 6y_3 &= 3 \end{aligned}$$

to get $y_1 = -10.833,\ y_2 = 5.333,\ y_3 = 6.833.$

Computing $A^t\mathbf{y}$ we get $x_1^* = -0.167$, $x_2^* = 1.333$, $x_3^* = -4.000$, $x_4^* = 2.833$ and finally $f(x_1^*,x_2^*,x_3^*,x_4^*) = 25.833$. □

3.5. EXERCISES

1. Show that the coefficient matrix of Example 4 has $\mathbb{R}^3$ for its column space.

2. Find the smallest value of $x_1^2 + x_2^2 + x_3^2 + x_4^2$ subject to $A\mathbf{x} = \mathbf{b}$ for the following A and $\mathbf{b}$.

(a) $A = \begin{bmatrix} 1 & 1 & 1 & 0 \\ 0 & 1 & 1 & 1 \end{bmatrix}$ $\mathbf{b} = \begin{bmatrix} 2 \\ 0 \end{bmatrix}$

(b) $A = \begin{bmatrix} 0 & 1 & 2 & 1 \\ 1 & 1 & 1 & 1 \end{bmatrix}$ $\mathbf{b} = \begin{bmatrix} 1 \\ 2 \end{bmatrix}$

(c) $A = \begin{bmatrix} 1 & 1 & 1 & 1 \\ 2 & 0 & 1 & 0 \\ 0 & 2 & 1 & 0 \end{bmatrix}$ $\mathbf{b} = \begin{bmatrix} 0 \\ 1 \\ 2 \end{bmatrix}$.

3. Find the function of the form $f(x) = a + bx^2 + cx^4$ that satisfies $f(0) = 2$, $f(1) = 4$, and minimizes $a^2 + b^2 + c^2$.

4. Let A be a 2×4 matrix with linearly independent rows. Define $A^\dagger = A^t(AA^t)^{-1}$. Verify that $AA^\dagger = I$ and that $A^\dagger AA^\dagger A = A^\dagger A$. Is $A^\dagger A = I$? Compare this to Exercise 9 of Section 3.2.

3.6. Areas, Volumes, and Determinants

We review the basic facts about 2×2 and 3×3 determinants and relate them to linear systems.

Consider the system

$$\begin{aligned} ax + by &= \alpha \\ cx + dy &= \beta \end{aligned} \qquad (1)$$

and assume $a \neq 0$. Then, multiplying the first equation by $\frac{c}{a}$ and subtracting the result from the second, we get

$$\begin{aligned} ax + by &= \alpha \\ (d - \tfrac{bc}{a})y &= \beta - \tfrac{c}{a}\alpha. \end{aligned}$$

Or,

$$\begin{aligned} ax + by &= \alpha \\ (ad - bc)y &= a\beta - c\alpha. \end{aligned}$$

So, as long as $$ad - bc \neq 0$$
we can solve for y and eventually for a, no matter what α and β were.

If it happened that $a = 0$, then we could interchange the order of the equations. If both $a = 0$ and $c = 0$, then there will be α and β for which the system has no solution. In summary we have

***Theorem* 1.** The system (1) has a unique solution for every α and β if and only if $ad - bc \neq 0$.

Definition. The *determinant* of $\begin{bmatrix} a & b \\ c & d \end{bmatrix}$ is $det\begin{bmatrix} a & b \\ c & d \end{bmatrix} = ad - bc$.

***Theorem* 2.** For $A = \begin{bmatrix} a & b \\ c & d \end{bmatrix}$ the following conditions are all equivalent.

(a) $det\ A \neq 0$

(b) the column space of A is $\mathbb{R}^2$.

(c) the null space of A is $\{0\}$

(d) the parallelogram determined by the columns of A has non zero area.

The equivalence of (a)-(c) follows from earlier work. For (d) we will show that $|det\ A|$ is the area of the parallelogram determined by the columns of A. First we note that if the columns are multiples of each other, then the parallelogram degenerates into a line segment which has zero area. On the other hand

$$det\begin{bmatrix} a & ta \\ c & tc \end{bmatrix} = tac - tac = 0.$$

Now, if the columns are not multiples of each other, they can be sketched in a picture like Figure 1.

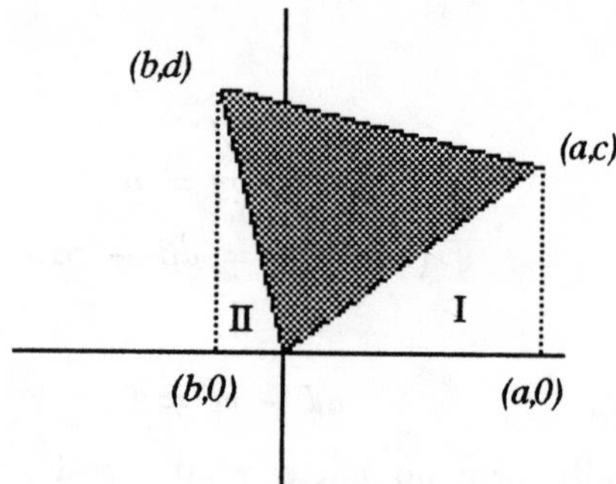

The trapezoid with vertices $(a,0)$, (a,c), (b,d), $(b,0)$ has area

$$\tfrac{1}{2}(a-b)(d+c) = \tfrac{1}{2}\{ad - bd + ac - bc\}.$$

Triangle I has area $\frac{1}{2}ac$. Triangle II has area $-\frac{1}{2}bd$.

So, the triangle determined by $\begin{bmatrix} a \\ c \end{bmatrix}$ and $\begin{bmatrix} b \\ d \end{bmatrix}$ has area

$$\tfrac{1}{2}\{ad - bd + ac - bc\} - \tfrac{1}{2}ac + \tfrac{1}{2}bd = \tfrac{1}{2}\{ad - bc\}.$$

Therefore, the parallelogram determined by the columns has area $ad - bc$.

The determinant of a 3×3 matrix with columns c_1, c_2 and c_3 can be defined by

$$det[c_1, c_2, c_3] = (c_1 \times c_2) \cdot c_3.$$

If c_2 is a multiple of c_1, then $c_1 \times c_2$ is the zero vector and the determinant is 0. If c_1 and c_2 determine a plane, then the normal to this plane is $c_1 \times c_2$ and c_3 is in this plane if and only if it is perpendicular to this normal. Thus, the column space of $[c_2\ \ c_2\ \ c_3]$ is $\mathbb{R}^3$ if and only if $(c_1 \times c_2)\cdot c_3 \neq 0$. In Calculus it is shown in fact that the volume of the parallelopiped determined by c_1, c_2, and c_3 is $|det[c_1\ \ c_2\ \ c_3]|$ (cf. [5, p. 651]). In general we have

Theorem 3. Let A be a 3×3 matrix. The following conditions are all equivalent.

(a) $det\ A \neq 0$

(b) the column space of A is $\mathbb{R}^3$.

(c) the null space of A is $\{0\}$

(d) the columns of A determine a box (parallelopiped) with positive volume, $|det\ A|$.

There is a notion of determinant for $n \times n$ matrices. We will not discuss it, but simply note that the geometric interpretation relates the n-dimensional volume of the parallelopiped determined by the columns to the absolute value of the determinant.

Example 1. Find the area of the parallelogram determined by the columns of

$$\begin{bmatrix} \sqrt{2} & 3 \\ -1 & -4 \end{bmatrix}$$

Solution. The determinant is $-4\sqrt{2} + 3$ which is negative. So, the area is $4\sqrt{2} - 3$. □

Example 2. Find the area of the parallelogram whose vertices are (1,2), (0,5) and (−1,3), and (0,0).

Solution.

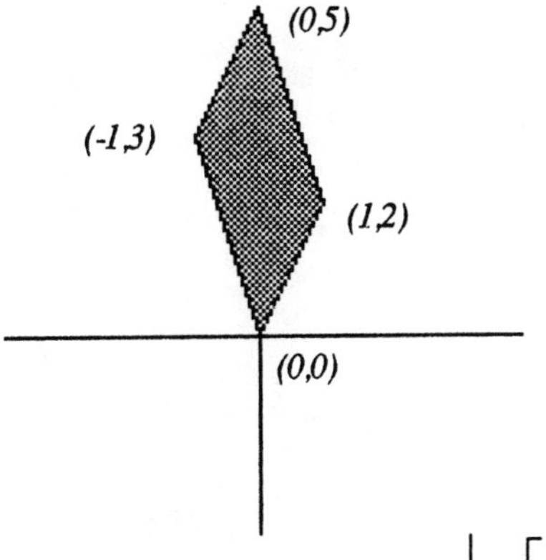

Since $(5,0) = (1,2) + (-1,3)$, the area is $\left| det \begin{bmatrix} 1 & -1 \\ 2 & 3 \end{bmatrix} \right|$ which is 5. □

3.6. EXERCISES

1. **Find the areas of the parallelograms whose vertices are given**

   (a) (1,2), (0,0), (4,3), (5,5)

   (b) (0,0), (2,2), (5,−1), (7,1)

   (c) (0,0), (−1,−1), (−1,1), (−2,0).

2. **Find the volume of the parallelepiped determined by the columns of** $\begin{bmatrix} 1 & 1 & 0 \\ 2 & 0 & 1 \\ 3 & 1 & 0 \end{bmatrix}$.

3. **Use determinants to show that there exist b so that** $A\mathbf{x} = \mathbf{b}$ **has no solution if**

   (a) $A = \begin{bmatrix} 1 & 2 \\ 2 & 4 \end{bmatrix}$

   (b) $A = \begin{bmatrix} 1 & -1 \\ -1 & 1 \end{bmatrix}$

   (c) $A = \begin{bmatrix} 1 & 1 & 3 \\ 2 & 1 & 5 \\ 3 & 0 & 6 \end{bmatrix}$

   (d) $A = \begin{bmatrix} 1 & 3 & 1 \\ 2 & 0 & 1 \\ -2 & 6 & 0 \end{bmatrix}$

4. Let $A = \begin{bmatrix} a & b & c \\ d & e & f \\ g & h & i \end{bmatrix}$. Show that $det\ A = aei + bfg + cdh - gec - hfa - idb.$

CHAPTER 4 Linear Functions and Matrices

The coefficient matrix of a system $A\mathbf{x} = \mathbf{b}$ of n equations in k unknowns has a natural interpretation as a function from $\mathbb{R}^k$ to $\mathbb{R}^n$. The vector $\mathbf{x}$ in $\mathbb{R}^k$ is transformed into the vector $\mathbf{b}$ in $\mathbb{R}^n$. Some important applications of matrices are concerned with the geometric aspects of this interpretation. In this chapter we study the geometric properties 2×2 and 3×3 matrices viewed as transformations.

4.1. Basic Properties of Matrix Functions

We recall the column oriented approach to matrix multiplication: the product $A\mathbf{x}$ is a weighted sum of the column vectors of A. This has several important consequences:

(1) if we add two vectors and then apply A to the sum, we get the same result that we'd get if we applied A to each vector and added the results. Symbolically, for all vectors $\mathbf{x}$ and $\mathbf{y}$ we have

$$A(\mathbf{x} + \mathbf{y}) = A\mathbf{x} + A\mathbf{y}.$$

(2) if we multiply a vector $\mathbf{x}$ by a number s and then apply A, we get the same result that we'd get by applying A to the vector and multiplying the result by s, i.e., for all vectors $\mathbf{x}$ and for all numbers s we have

$$A(s\mathbf{x}) = s(A\mathbf{x}).$$

To see (1) let the columns of A be $\mathbf{a}_1,\mathbf{a}_2,\ldots,\mathbf{a}_k$. Then,

$$A\mathbf{x} = x_1\mathbf{a}_1 + x_2\mathbf{a}_2 + \cdots + x_k\mathbf{a}_k \text{ and } A\mathbf{y} = y_1\mathbf{a}_1 + y_2\mathbf{a}_2 + \cdots + y_k\mathbf{a}_k;$$

so,

$$A\mathbf{x} + A\mathbf{y} = (x_1+y_1)\mathbf{a}_1 + (x_2+y_2)\mathbf{a}_2 + \cdots + (x_k+y_k)\mathbf{a}_k = A(\mathbf{x} + \mathbf{y}).$$

Similarly, if s is a real number, then

$$sA\mathbf{x} = sx_1\mathbf{a}_1 + \cdots + sx_n\mathbf{a}_n = A(s\mathbf{x})$$

From (1) and (2) we can deduce an important geometric property of matrix induced functions — the line segment joining two points $\mathbf{p}$ and $\mathbf{q}$ in $\mathbb{R}^k$ is taken to the line segment joining the points $A\mathbf{p}$ and $A\mathbf{q}$ in $\mathbb{R}^n$. To see this first recall that points on the line segment joining $\mathbf{v}$ and $\mathbf{u}$ are all those points of the form $t\mathbf{v} + (1-t)\mathbf{u}$ for $0 \le t \le 1$. {For example, the midpoint is obtained by taking $t = \frac{1}{2}$.} Now consider a general point on the segment joining $\mathbf{p}$ and $\mathbf{q}$ say $t\mathbf{p} + (1-t)\mathbf{q}$. Then

$$\begin{aligned} A(t\mathbf{p} + (1-t)\mathbf{q}) &= A(t\mathbf{p}) + A((1-t)\mathbf{q}) && \text{by (1)} \\ &= tA\mathbf{p} + (1-t)A\mathbf{q} && \text{by (2).} \end{aligned}$$

Since $tA\mathbf{p} + (1-t)A\mathbf{q}$ for $0 \le t \le 1$ is a typical point on the segment joining $A\mathbf{p}$ and $A\mathbf{q}$, we have that the line between $\mathbf{p}$ and $\mathbf{q}$ goes onto the line between $A\mathbf{p}$ and $A\mathbf{q}$. In fact even more can be said. Taking $t = \frac{1}{2}$ we see that midpoints go to midpoints, with $t = \frac{1}{3}$ we see that the point one-third of the way from $\mathbf{q}$ to $\mathbf{p}$ is sent to the point one-third of the way from $A\mathbf{q}$ to $A\mathbf{p}$, and so on. In summary we have the following.

Theorem 1. The mapping from $\mathbb{R}^k$ to $\mathbb{R}^n$ given by the $n{\times}k$ matrix A takes line segments to line segments and preserves proportions.

Example 1. Let $A = \begin{bmatrix} 3 & 1 \\ 1 & 2 \end{bmatrix}$ and $\mathbf{p} = (1,1)$ and $\mathbf{q} = (-1,1)$. Sketch and label the points $t\mathbf{p} + (1-t)\mathbf{q}$ and $A(t\mathbf{p} + (1-t)\mathbf{q})$ for the following values of t: $\frac{1}{2}, \frac{1}{4}, \frac{3}{4}, \frac{1}{3}, \frac{4}{5}$.

Solution.

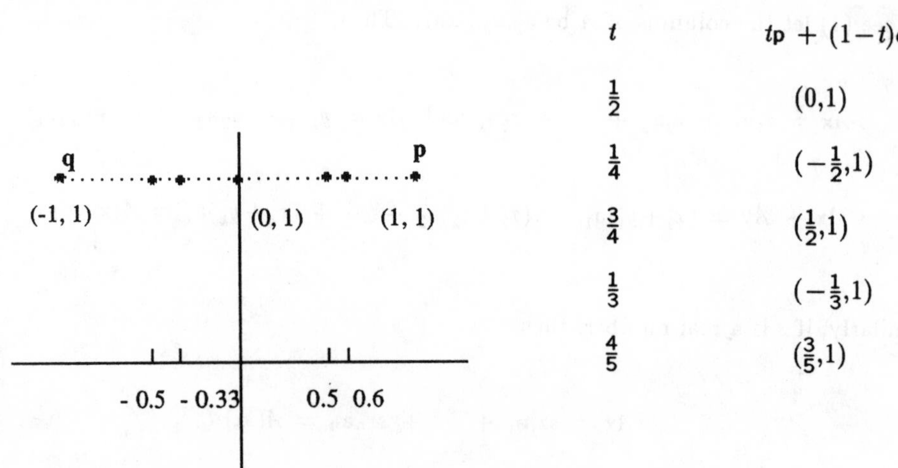

t	$t\mathbf{p} + (1-t)\mathbf{q}$
$\frac{1}{2}$	$(0,1)$
$\frac{1}{4}$	$(-\frac{1}{2},1)$
$\frac{3}{4}$	$(\frac{1}{2},1)$
$\frac{1}{3}$	$(-\frac{1}{3},1)$
$\frac{4}{5}$	$(\frac{3}{5},1)$

Now, $A\mathbf{p} = \begin{bmatrix} 3 & 1 \\ 1 & 2 \end{bmatrix}\begin{bmatrix} 1 \\ 1 \end{bmatrix} = \begin{bmatrix} 4 \\ 3 \end{bmatrix}$ and $A\mathbf{q} = \begin{bmatrix} -2 \\ 1 \end{bmatrix}$ so the points $A(t\mathbf{p} + (1-t)\mathbf{q})$ should be on the line segment joining (4,3) and (−2,1).

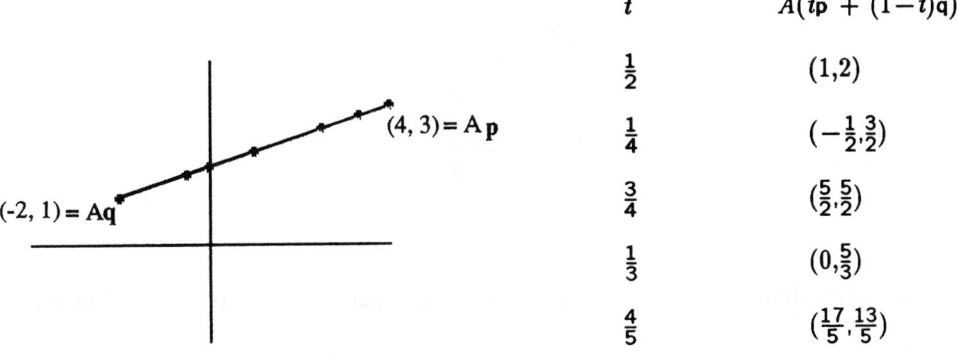

t	$A(t\mathbf{p} + (1-t)\mathbf{q})$
$\frac{1}{2}$	(1,2)
$\frac{1}{4}$	$(-\frac{1}{2},\frac{3}{2})$
$\frac{3}{4}$	$(\frac{5}{2},\frac{5}{2})$
$\frac{1}{3}$	$(0,\frac{5}{3})$
$\frac{4}{5}$	$(\frac{17}{5},\frac{13}{5})$

□

Definition. The *image of a set S under a matrix A* is the set into which A transforms S, that is, image of $S = \{Ax\colon x \in S\}$.

Example 2. Sketch the image of the triangle with vertices (1,2), (−2,1), and (3,−2) under the matrix

$$M = \begin{bmatrix} 0 & 1 \\ -1 & 0 \end{bmatrix}.$$

Solution. We have

$$M\begin{bmatrix} 1 \\ 2 \end{bmatrix} = \begin{bmatrix} 2 \\ -1 \end{bmatrix}, \quad M\begin{bmatrix} -2 \\ 1 \end{bmatrix} = \begin{bmatrix} 1 \\ 2 \end{bmatrix}, \quad M\begin{bmatrix} 3 \\ -2 \end{bmatrix} = \begin{bmatrix} -2 \\ -3 \end{bmatrix}$$

so the given triangle goes to the triangle with vertices (2,−1), (1,2), (−2,−3).

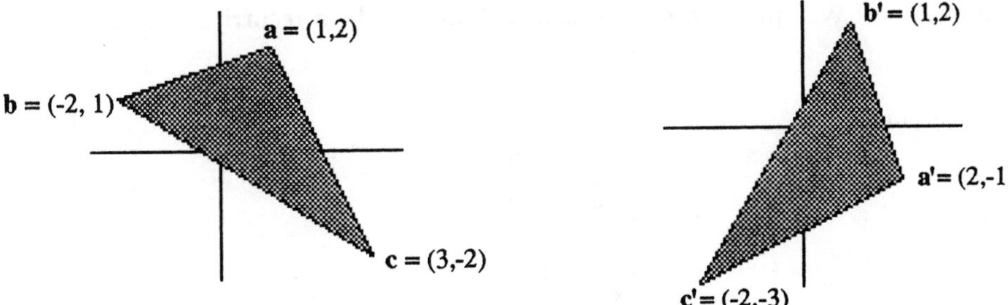

□

$M\mathbf{a} = \mathbf{a}'$, $M\mathbf{b} = \mathbf{b}'$, $M\mathbf{c} = \mathbf{c}'$

Example 3. Find the image of the pentagon sketched below under the matrix $M = \begin{bmatrix} 2 & 0 \\ 0 & -1 \end{bmatrix}$.

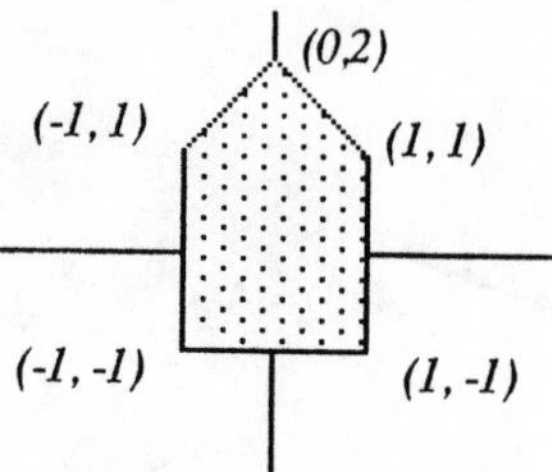

Solution. We must find the image of each corner of the pentagon. This will tell us how the sides get transformed. We have

$$M\begin{bmatrix} 0 \\ 2 \end{bmatrix} = \begin{bmatrix} 0 \\ -2 \end{bmatrix}, \quad M\begin{bmatrix} -1 \\ 1 \end{bmatrix} = \begin{bmatrix} -2 \\ -1 \end{bmatrix}, \quad M\begin{bmatrix} -1 \\ -1 \end{bmatrix} = \begin{bmatrix} -2 \\ 1 \end{bmatrix}, \quad M\begin{bmatrix} 1 \\ -1 \end{bmatrix} = \begin{bmatrix} 2 \\ 1 \end{bmatrix}, \quad M\begin{bmatrix} 1 \\ 1 \end{bmatrix} = \begin{bmatrix} 2 \\ -1 \end{bmatrix}.$$

So, the image is

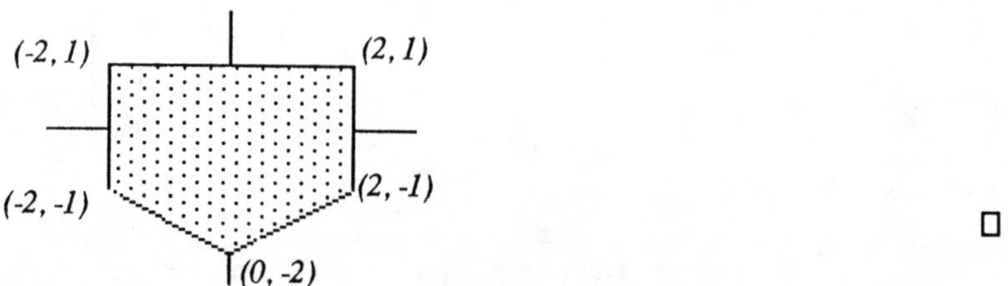

□

Remark. From the above examples we see that matrices take polygons to polygons and cannot increase the number of vertices. However, it can be decreased as we'll see in the next example. Finally, note that in the above examples we have applied a matrix to a set; this will be a recurring theme. □

Example 4. Show that the matrix

$$M = \begin{bmatrix} \frac{1}{2} & \frac{1}{2} \\ \frac{1}{2} & \frac{1}{2} \end{bmatrix}$$

takes the triangle of Example 2 to a line segment.

Solution. The column space of M is the line $y = x$, so M takes all $\mathbb{R}^2$ to a line. The triangle of Example 2 must go into a subset of this line. Since

$$M\begin{bmatrix} 1 \\ 2 \end{bmatrix} = \begin{bmatrix} \frac{3}{2} \\ \frac{3}{2} \end{bmatrix}, \quad M\begin{bmatrix} -2 \\ 1 \end{bmatrix} = \begin{bmatrix} -\frac{1}{2} \\ -\frac{1}{2} \end{bmatrix}, \text{ and } M\begin{bmatrix} 3 \\ -2 \end{bmatrix} = \begin{bmatrix} \frac{1}{2} \\ \frac{1}{2} \end{bmatrix},$$

we see that the required line segment is the straight line between $(-\frac{1}{2},-\frac{1}{2})$ and $(\frac{3}{2},\frac{3}{2})$. □

What does a 2×2 matrix do to circles? If the column space is a line, then a circle is taken to a line segment. In the general case when the column space is all of $\mathbb{R}^2$, circles go into ellipses. We will not attempt to justify this in general. But consider the following example.

Example 5. Show that the matrix $D = \begin{bmatrix} 2 & 0 \\ 0 & -4 \end{bmatrix}$ takes the circle $x^2 + y^2 = R^2$ into an ellipse.

Solution. (x,y) is on the given circle if and only if $x = R \cos t$, $y = R \sin t$ for $0 \le t \le 2\pi$. Then, if

$$\begin{bmatrix} u \\ v \end{bmatrix} = D\begin{bmatrix} x \\ y \end{bmatrix},$$

we have $u = 2R \cos t$, $v = -4\ R \sin t$ so

$$(\tfrac{u}{2})^2 + (\tfrac{v}{4})^2 = R^2 \cos^2 t + R^2 \sin^2 t = R^2$$

which is the equation of an ellipse.

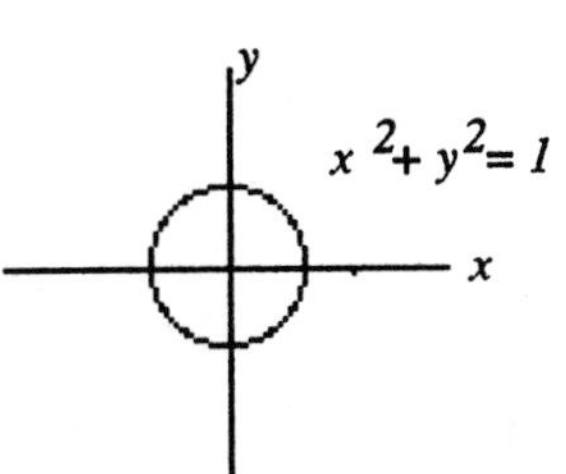

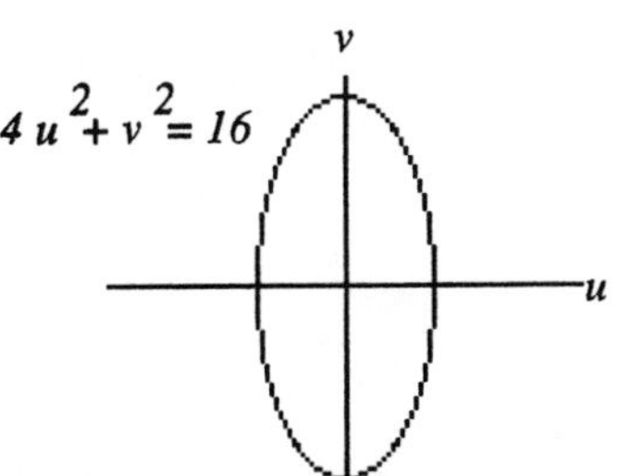

□

4.1. EXERCISES

1. Let $\mathbf{p} = (1,0)$ and $\mathbf{q} = (0,1)$. Find the image of the line segment between $\mathbf{p}$ and $\mathbf{q}$ under the following matrices. In each case sketch the image and label the images of $t\mathbf{p} + (1-t)\mathbf{q}$ for $t = \frac{1}{4}, \frac{1}{2}, \frac{3}{4}$.

(a) $\begin{bmatrix} 2 & 1 \\ 1 & 2 \end{bmatrix}$ (b) $\begin{bmatrix} 0 & 1 \\ -1 & 0 \end{bmatrix}$ (c) $\begin{bmatrix} -4 & 0 \\ 0 & 2 \end{bmatrix}$

(d) $\begin{bmatrix} 2 & 0 \\ 0 & -3 \end{bmatrix}$ (e) $\begin{bmatrix} \frac{1}{2} & \frac{1}{2} \\ \frac{1}{2} & \frac{1}{2} \end{bmatrix}$ (f) $\begin{bmatrix} 1 & 1 \\ -1 & -1 \end{bmatrix}$

2. Sketch the image of the square with vertices (0,0), (0,1), (1,1), (1,0) under the following matrices.

(a) $\begin{bmatrix} \frac{\sqrt{2}}{2} & -\frac{\sqrt{2}}{2} \\ \frac{\sqrt{2}}{2} & \frac{\sqrt{2}}{2} \end{bmatrix}$ (b) $\begin{bmatrix} 1 & 0 \\ 0 & -2 \end{bmatrix}$ (c) $\begin{bmatrix} 0 & 2 \\ 1 & 0 \end{bmatrix}$

(d) $\begin{bmatrix} 2 & 1 \\ 1 & 2 \end{bmatrix}$ (e) $\begin{bmatrix} 1 & 0 \\ 0 & 0 \end{bmatrix}$ (f) $\begin{bmatrix} 1 & 1 \\ -1 & 1 \end{bmatrix}$

3. Sketch the image of the circle $x^2 + y^2 = 1$ under the matrices (b), (c), (e) of Exercise 1.

4.2. Linear Functions

The properties (1) and (2) of matrix functions can be axiomatized.

Definition. A function $f: \mathbb{R}^k \to \mathbb{R}^n$ is called *linear* if for every pair of vectors $\mathbf{x}$ and $\mathbf{y}$ in $\mathbb{R}^k$ and for every real number s we have

$$f(\mathbf{x} + \mathbf{y}) = f(\mathbf{x}) + f(\mathbf{y}) \qquad (1)$$

$$f(s\mathbf{x}) = sf(\mathbf{x}). \qquad (2)$$

One can sometimes show that certain geometrically defined functions are linear functions. For example rigid motions of the plane which keep the origin fixed are linear because they preserve straight lines and angles.

The most important fact about linear functions is that all linear functions from $\mathbb{R}^k$ to $\mathbb{R}^n$ are determined by matrices. To be precise we have

Theorem 1. If $f: \mathbb{R}^k \to \mathbb{R}^n$ is a linear function, then there is an $n \times k$ matrix M so that $f(\mathbf{x}) = M\mathbf{x}$ for all $\mathbf{x}$ in $\mathbb{R}^k$. M is said to *represent* f.

So if you know that a particular function is linear, how can you find the matrix that represents it? The answer is a consequence of Theorem 1 of Section 2.1, which we recall as

Theorem 2. If A is an $n \times k$ matrix, then the columns of A are the vectors $A\mathbf{e}_1, A\mathbf{e}_2, \ldots, A\mathbf{e}_k$ where $\mathbf{e}_1, \ldots, \mathbf{e}_k$ are the standard unit vectors in $\mathbb{R}^k$.

Now, if f is known to be linear, then the ith column of the matrix M that represents f is $M\mathbf{e}_i = f(\mathbf{e}_i)$.

Corollary. If f is a linear function from $\mathbb{R}^k$ to $\mathbb{R}^n$, then the matrix that represents f is

$$M_f = [f(\mathbf{e}_1) \quad f(\mathbf{e}_2) \quad \cdots \quad f(\mathbf{e}_k)].$$

Example 1. Show that the only matrix with the property $A\mathbf{x} = \mathbf{b}$ for *all* $x \in \mathbb{R}$ is the identity matrix.

Solution. $A\mathbf{e}_1 = \mathbf{e}_1$ so the first column is $\begin{bmatrix} 1 \\ 0 \end{bmatrix}$. Similarly $A\mathbf{e}_2 = \mathbf{e}_2 = \begin{bmatrix} 0 \\ 1 \end{bmatrix}$ is the second column. □

Theorem 2 and its corollary are the main tools used in the applications in the rest of this chapter.

Scalings

The simplest class of linear functions from $\mathbb{R}^k$ to $\mathbb{R}^n$ are those which scale the coordinate axes. They have the general form

$$f(x_1,x_2,\ldots,x_n) = (s_1x_1,s_2x_2,\ldots,s_nx_n).$$

The numbers $s_1,s_2,\ldots,s_n$ are the *scale factors.* The matrix that represents a scaling has the form of a *diagonal matrix.*

We consider the 2×2 case in detail. If $f(x_1,x_2) = (s_1x_1,s_2x_2)$, then

$$f(\mathbf{e}_1) = f(1,0) = (s_1,0) = s_1\mathbf{e}_1$$

and

$$f(\mathbf{e}_2) = f(0,1) = (0,s_1) = s_2\mathbf{e}_2.$$

So, the matrix that represents f is

$$\begin{bmatrix} s_1 & 0 \\ 0 & s_2 \end{bmatrix}.$$

Negative values of s_1 and s_2 correspond to reflections of the coordinate axes.

Example 2. Sketch the image of the square determined by $\mathbf{e}_1$ and $\mathbf{e}_2$ under the scaling function $f(x_1,x_2) = (3x_1,-2x_2)$.

Solution. We note in passing that the matrix that represents f is

$$D = \begin{bmatrix} 3 & 0 \\ 0 & -2 \end{bmatrix}.$$

Now, $D\mathbf{e}_1 = 3\mathbf{e}_1$ and $D\mathbf{e}_2 = -2\mathbf{e}_2$ gives:

Example 3. The function f that rotates the plane about the origin by $\frac{\pi}{2}$ radians in a counterclockwise manner is known to be linear. Find the matrix that represents this function and find the image of the triangle whose vertices are (1,1), (2,1), (2,3) under it.

Solution. Let M be the desired matrix. The first column of M is $f(\mathbf{e}_1)$ and the second is $f(\mathbf{e}_2)$:

$$M = [f(\mathbf{e}_1) \quad f(\mathbf{e}_2)].$$

It is clear that $f(\mathbf{e}_1) = \mathbf{e}_2$ and $f(\mathbf{e}_2) = -\mathbf{e}_1$:

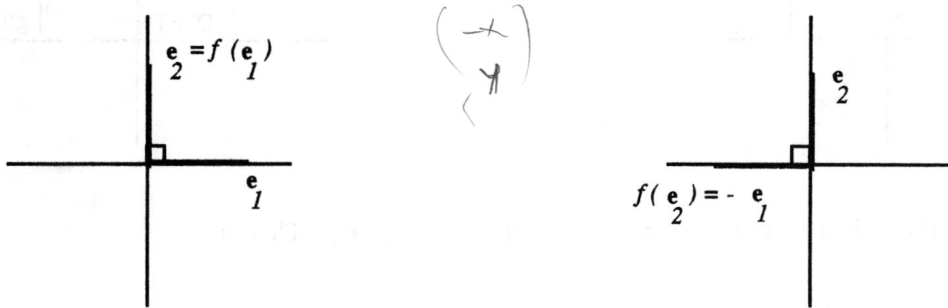

Therefore,

$$M = \begin{bmatrix} 0 & -1 \\ 1 & 0 \end{bmatrix}.$$

Now,

$$M\begin{bmatrix}1\\1\end{bmatrix}=\begin{bmatrix}-1\\1\end{bmatrix}\quad M\begin{bmatrix}2\\1\end{bmatrix}=\begin{bmatrix}-1\\2\end{bmatrix}\quad M\begin{bmatrix}2\\3\end{bmatrix}=\begin{bmatrix}-3\\2\end{bmatrix}.$$

So, M takes the given triangle onto the triangle whose vertices are $(-1,1)$, $(-1,2)$ and $(-3,2)$.

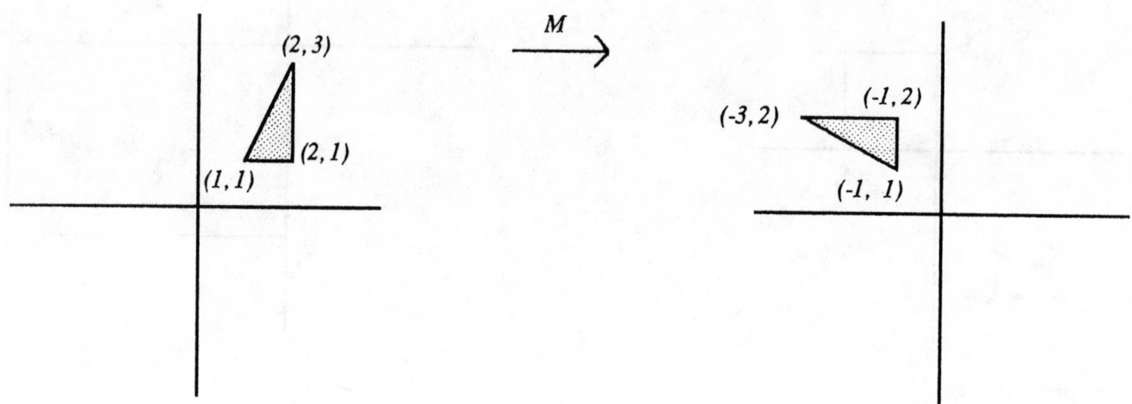

□

Example 4. Let g be the function that reflects (flips) the plane through the line $y = x$. It is known that g is linear. Find the matrix that represents g and find the image under g of the triangle given in Example 2.

Solution. g turns the positive x-axis into the positive y-axis and the positive y-axis into the positive x-axis:

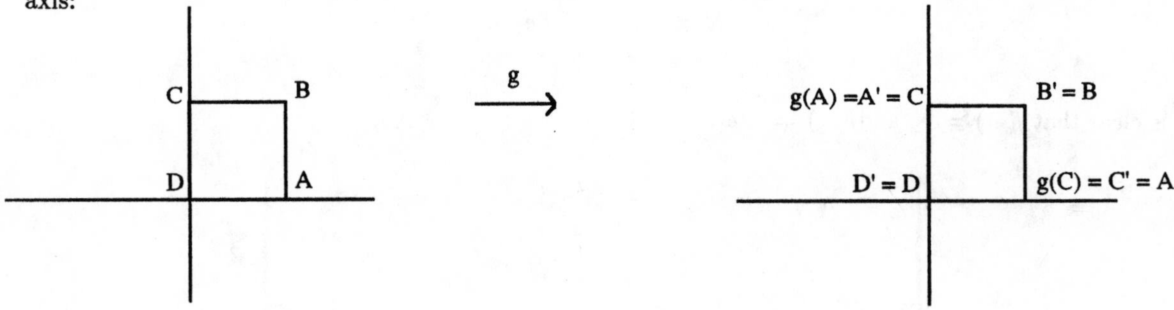

Let F be the desired matrix, so $F = [g(\mathbf{e}_1) \quad g(\mathbf{e}_2)] = [\mathbf{e}_2 \quad \mathbf{e}_1]$. That is

$$F = \begin{bmatrix}0 & 1\\1 & 0\end{bmatrix}$$

Now,

$$F\begin{bmatrix}1\\1\end{bmatrix}=\begin{bmatrix}1\\1\end{bmatrix},\quad F\begin{bmatrix}2\\1\end{bmatrix}=\begin{bmatrix}1\\2\end{bmatrix}\quad\text{and}\quad F\begin{bmatrix}2\\3\end{bmatrix}=\begin{bmatrix}3\\2\end{bmatrix}.$$

So, we have

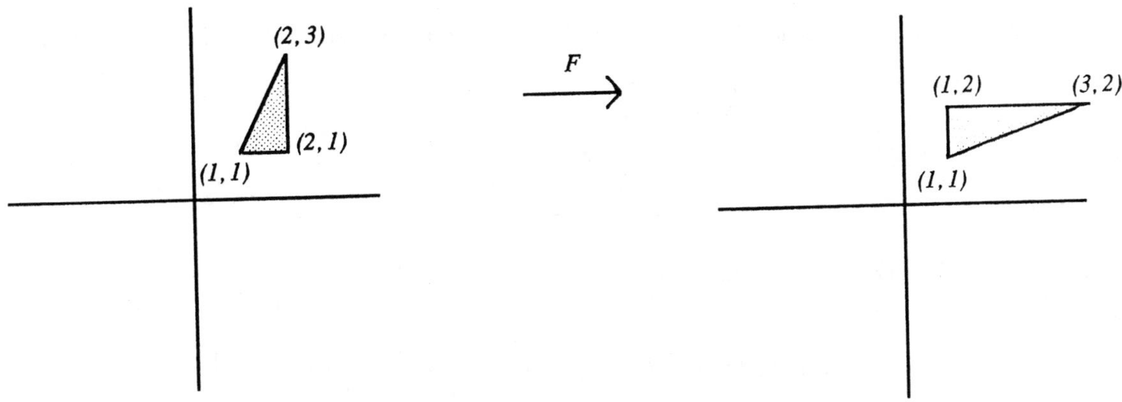

□

Example 5. Find the 3×3 matrix that represents the linear function f which rearranges coordinates by the rule $f(x_1,x_2,x_3) = (x_2,x_3,x_1)$.

Solution. The desired matrix is

$$B = [f(\mathbf{e}_1) \quad f(\mathbf{e}_2) \quad f(\mathbf{e}_3)].$$

Now $f(\mathbf{e}_1) = f(1,0,0) = (0,0,1)$, $f(\mathbf{e}_2) = (1,0,0)$, and $f(\mathbf{e}_3) = (0,1,0)$. So,

$$B = \begin{bmatrix}0 & 1 & 0\\0 & 0 & 1\\1 & 0 & 0\end{bmatrix}.$$

To check this notice

$$B\begin{bmatrix}x_1\\x_2\\x_3\end{bmatrix} = \begin{bmatrix}0 & 1 & 0\\0 & 0 & 1\\1 & 0 & 0\end{bmatrix}\begin{bmatrix}x_1\\x_2\\x_3\end{bmatrix} = \begin{bmatrix}x_2\\x_3\\x_1\end{bmatrix}.$$

□

In some applications f is linear but the quantities $f(\mathbf{e}_k)$ are not explicitly given and an intermediate computation must be done.

Example **6.** Find the 2×2 matrix that represents the linear function $g(\mathbf{x})$ that satisfies $g(1,1) = (1,1)$ and $g(-1,1) = (-3,3)$.

Solution. We use the fact that g is linear to find $g(1,0)$ and $g(0,1)$ in terms of $g(1,1)$ and $g(-1,1)$. Since $(1,1) + (-1,1) = (0,2) = 2\mathbf{e}_2$ we have

$$\mathbf{e}_2 = \frac{1}{2}\begin{bmatrix} 1 \\ 1 \end{bmatrix} + \frac{1}{2}\begin{bmatrix} -1 \\ 1 \end{bmatrix}.$$

Using the properties (1) and (2) of linearity we get

$$g(\mathbf{e}_2) = \frac{1}{2}\, g(1,1) + \frac{1}{2} g(-1,1) = \frac{1}{2}\begin{bmatrix} 1 \\ 1 \end{bmatrix} + \frac{1}{2}\begin{bmatrix} -3 \\ 3 \end{bmatrix} = \begin{bmatrix} -1 \\ 2 \end{bmatrix}$$

which is the second column of our matrix. Similarly,

$$\mathbf{e}_1 = \frac{1}{2}\begin{bmatrix} 1 \\ 1 \end{bmatrix} - \frac{1}{2}\begin{bmatrix} -1 \\ 1 \end{bmatrix}$$

so

$$g(\mathbf{e}_1) = \frac{1}{2}\, g(1,1) - \frac{1}{2} g(-1,1) = \frac{1}{2}\begin{bmatrix} 1 \\ 1 \end{bmatrix} - \frac{1}{2}\begin{bmatrix} -3 \\ 3 \end{bmatrix} = \begin{bmatrix} 2 \\ -1 \end{bmatrix}$$

Hence, the required matrix is $\begin{bmatrix} 2 & -1 \\ -1 & 2 \end{bmatrix}$. □

Compositions

If f is a linear function from $\mathbb{R}^n$ to $\mathbb{R}^k$ and g is a linear function from $\mathbb{R}^k$ to $\mathbb{R}^p$, then the *composition* $g \circ f$ is the function from $\mathbb{R}^n$ to $\mathbb{R}^p$ defined by

$$(g \circ f)(\mathbf{x}) = g(f(\mathbf{x})).$$

It can be shown that $g \circ f$ is also linear. The matrix that represents $g \circ f$ can be easily found if the matrices that represent g and f are known.

Theorem 3. Let f and g be as above and let M_f and M_g be the matrices that represent f and g respectively. Then, the matrix that represents $g \circ f$ is

$$M_{g \circ f} = M_g M_f.$$

To see this consider the first column of $M_{g \circ f}$. This must be $g(f(\mathbf{e}_1))$ which is g applied to the first column of M_f. But for any vector $\mathbf{x}$ we have $g(\mathbf{x}) = M_g(\mathbf{x})$. In particular $g(f(\mathbf{e}_1)) = M_g(f(\mathbf{e}_1))$. So, we have

$$M_{g \circ f}(\mathbf{e}_1) = g(f(\mathbf{e}_1)) = M_g(f(\mathbf{e}_1)).$$

We have similar statements for every column of $M_{g \circ f}$:

$$M_{g \circ f}(\mathbf{e}_i) = g(f(\mathbf{e}_i)) = M_g(f(\mathbf{e}_i)) \quad i = 1,2,\ldots,n.$$

Therefore, writing these equations in matrix form we have

$$M_{g \circ f} = [M_{g \circ f}(\mathbf{e}_1) \quad M_{g \circ f}(\mathbf{e}_2) \quad \cdots \quad M_{g \circ f}(\mathbf{e}_n)] = [M_g(f(\mathbf{e}_1)) \quad M_g(f(\mathbf{e}_2)) \quad \cdots \quad M_g(f(\mathbf{e}_n))].$$

$$= M_g[f(\mathbf{e}_1) \quad f(\mathbf{e}_2) \quad \cdots \quad f(\mathbf{e}_n))], \text{ by the definition of matrix multiplication.}$$

$$= M_g M_f \text{ by the definition of } M_f.$$ □

Example 7. Let f and g take $\mathbb{R}^2$ to $\mathbb{R}^2$ be the linear functions defined as follows. f rotates the plane 90° counterclockwise about the origin and g doubles lengths on the x-axis and halves lengths on the y-axis: $g(x,y) = (2x, \frac{1}{2}y)$. Find the matrices that represent $g \circ f$ and $f \circ g$.

Solution. From Example 2 we know $M_f = \begin{bmatrix} 0 & -1 \\ 1 & 0 \end{bmatrix}$. From its definition, $g(\mathbf{e}_1) = 2\mathbf{e}_1$ and $g(\mathbf{e}_2) = \frac{1}{2}\mathbf{e}_2$ so $M_g = \begin{bmatrix} 2 & 0 \\ 0 & \frac{1}{2} \end{bmatrix}$. Now,

$$M_{g \circ f} = \begin{bmatrix} 2 & 0 \\ 0 & \frac{1}{2} \end{bmatrix} \begin{bmatrix} 0 & -1 \\ +1 & 0 \end{bmatrix} = \begin{bmatrix} 0 & 2 \\ -\frac{1}{2} & 0 \end{bmatrix}$$

and

$$M_{f\circ g} = \begin{bmatrix} 0 & 1 \\ -1 & 0 \end{bmatrix} \begin{bmatrix} 2 & 0 \\ 0 & \frac{1}{2} \end{bmatrix} = \begin{bmatrix} 0 & \frac{1}{2} \\ -2 & 0 \end{bmatrix}.$$

Here are pictures of the image of the square determined by $\mathbf{e}_1$ and $\mathbf{e}_2$ under each of these.

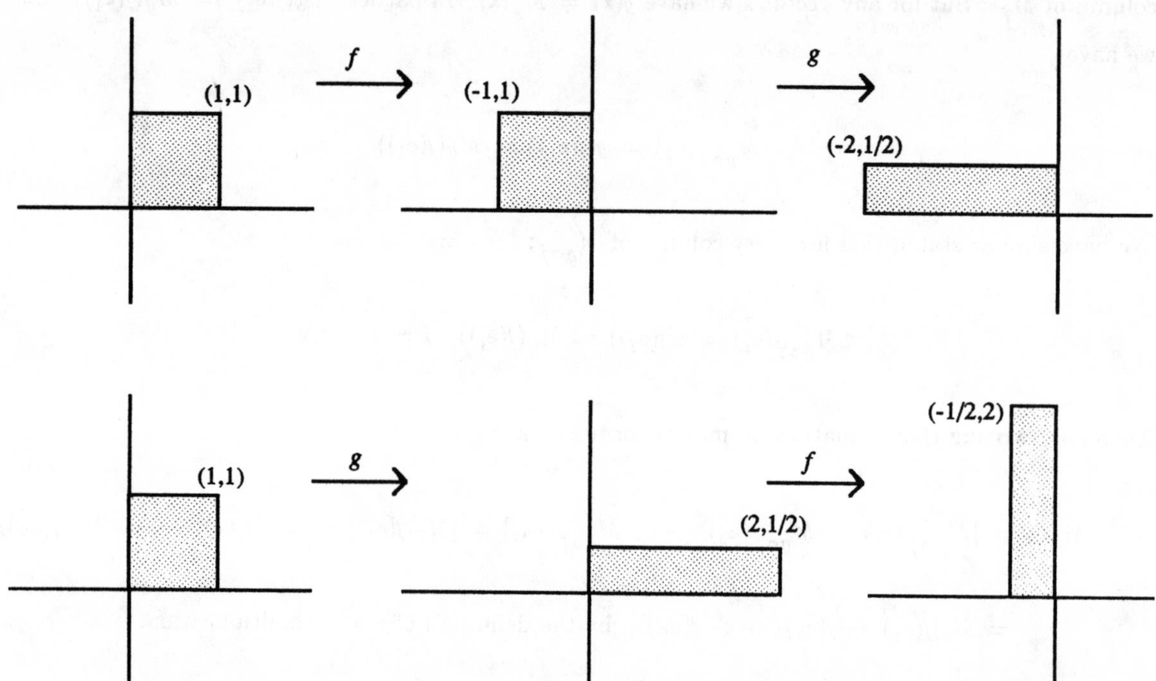

The pictures reinforce the fact that $M_{g\circ f} \neq M_{f\circ g}$. □

4.2. EXERCISES

1. **Find the 2×2 matrix that represents rotation of the plane 90° clockwise about the origin.**

2. **Find the 2×2 matrix that represents rotation of the plane 180° counterclockwise about the origin.**

3. Find a matrix that takes the x-axis to the line $y = x$ and the y-axis to the line $y = -x$. Can you arrange things so that the standard unit vectors are both sent to vectors of length 1? What form must the matrix have if the unit square is to go to a rectangle of area 1?

4. Suppose that f is a linear function from $\mathbb{R}^2$ to $\mathbb{R}^2$ that satisfies $f(1,1) = (-1,1)$ and $f(-1,1) = (-1,-1)$. Find the matrix that represents f and give a geometric interpretation of the action of f.

5. The process of projecting $\mathbb{R}^3$ onto the x-y plane is linear. Find the 3×3 matrix that represents this. The linear function is given by $f(x_1,x_2,x_3) = (x_1,x_2,0)$.

6. Let $F = \begin{bmatrix} 0 & 1 \\ 1 & 0 \end{bmatrix}$ and $M = \begin{bmatrix} 0 & -1 \\ 1 & 0 \end{bmatrix}$. Let $\mathcal{P}$ be the pentagon with vertices (0,2), (1,1), (1,−1), (−1,−1), (−1,1).
   (a) Sketch the images of $\mathcal{P}$ under F and M.
   (b) Sketch the images of $\mathcal{P}$ under each of the following. If possible, use your understanding of the geometric effects of F and M instead of algebraic computations. (Review Examples 3 and 4.)
   (i) FM (ii) MF (iii) F^2
   (iv) M^2 (v) MFM (vi) FMF.

7. Let $D = \begin{bmatrix} 3 & 0 \\ 0 & -2 \end{bmatrix}$ and let M and $\mathcal{P}$ be as in Exercise 5. Sketch the image of $\mathcal{P}$ under D, DM and MD.

8. Find the 3×3 matrix that scales the x_1-axis by 2, the x_2-axis by 3, and the x_3-axis by $\frac{1}{2}$. What is the volume of the box obtained by applying this matrix to the box determined by the standard unit vector?

9. Let $\mathbf{p}$ be a fixed number of $\mathbb{R}^3$. Use the Remark following Theorem 1 of Section 2.2 to show that the function defined by $\mathrm{F}(\mathbf{x}) = \mathbf{p}^t\mathbf{x}$ is a linear function from $\mathbb{R}^3$ to $\mathbb{R}$.

4.3. Translations and Affine Functions

It sometimes is necessary to select a new origin for our coordinate system. Translation functions facilitate this. Compositions of linear functions and translations define an important class of functions called affine functions.

Definitions. 1. A function from $\mathbb{R}^n$ to $\mathbb{R}^n$ of the form $f(\mathbf{x}) = \mathbf{x} + \mathbf{c}$ where $\mathbf{c}$ is a fixed vector in $\mathbb{R}^n$ is called a *translation*.

2. An *affine function* from $\mathbb{R}^k$ to $\mathbb{R}^n$ is a function of the form $f(\mathbf{x}) = M\mathbf{x} + \mathbf{c}$ where M is an $n \times k$ matrix and $\mathbf{c}$ is a fixed vector in $\mathbb{R}^n$.

Since translations preserve line segments and proportions we have the following extension of Theorem 1 of Section 4.1.

Theorem 1. If f is an affine function from $\mathbb{R}^k$ to $\mathbb{R}^n$, then f preserves line segments and proportions.

Example 1. The function $T\mathbf{x} = \mathbf{x} + \begin{bmatrix} 1 \\ 1 \end{bmatrix}$ translates the plane one unit to the right and one unit up. Sketch the image of the pentagon $\mathcal{P}$ of Exercise 5 of Section 4.1 under T.

Solution. We add $\begin{bmatrix} 1 \\ 1 \end{bmatrix}$ to each vertex and draw the sides:

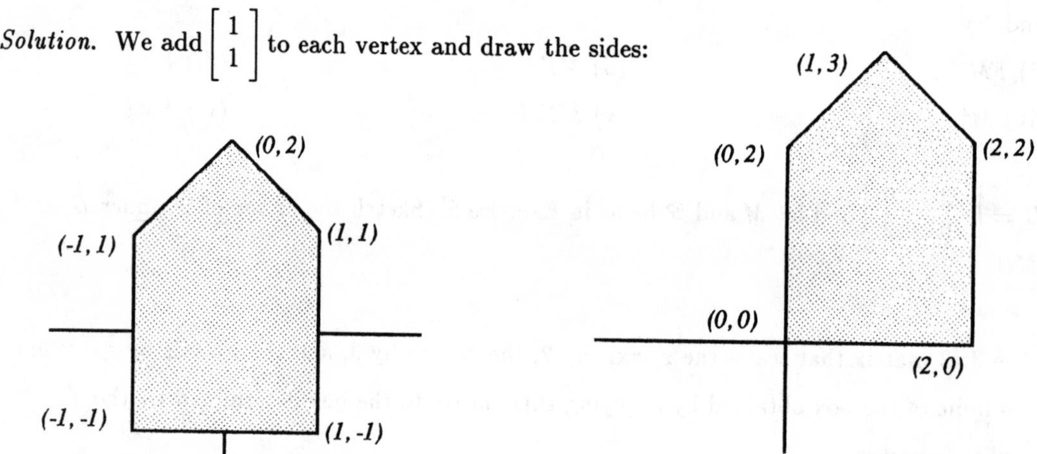

□

Example 2. Let T be as in Example 1 and let $M = \begin{bmatrix} 0 & -1 \\ 1 & 0 \end{bmatrix}$. Find the image of the pentagon in Example 1 under each of the following functions: f_1 is defined by first mapping T and then applying M to the result; f_2 is defined by first mapping M and then applying T to the result.

Solution. For f_1:

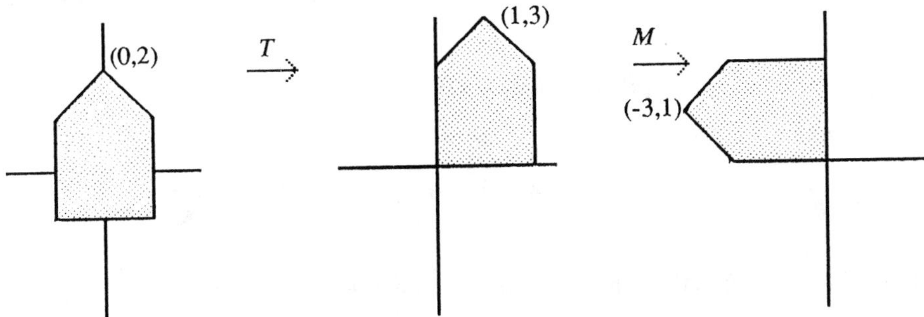

For f_2:

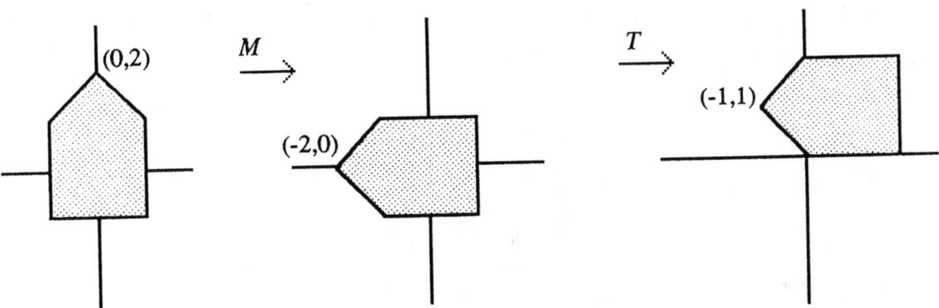

□

The preceding example shows that the order in which translations and matrices are applied to objects is important in determining their images. We could have seen this algebraically by computing the formulas for f_1 and f_2:

$$f_1(\mathbf{x}) = M(T\mathbf{x}) = M\left(\mathbf{x} + \begin{bmatrix} 1 \\ 1 \end{bmatrix}\right) = M\mathbf{x} + M\begin{bmatrix} 1 \\ 1 \end{bmatrix} = M\mathbf{x} + \begin{bmatrix} -1 \\ 1 \end{bmatrix}$$

$$f_2(\mathbf{x}) = T(M\mathbf{x}) = M\mathbf{x} + \begin{bmatrix} 1 \\ 1 \end{bmatrix}.$$

The matrix M in the above example rotates the plane 90° in a counterclockwise manner about the origin. If we want to rotate about a different point, $\mathbf{p}$, we use the function $f(\mathbf{x}) = M(\mathbf{x}-\mathbf{p}) + \mathbf{p}$. The geometric effect of f can be broken down into three simple actions:

First, translate so that $\mathbf{p}$ goes to 0; this is the translation $T_1\mathbf{x} = \mathbf{x} - \mathbf{p}$.

Second, rotate the plane 90° counterclockwise; this is where M comes in.

Third, translate so that **0** goes to p; this is the transation $T_2\mathbf{x} = \mathbf{x} + \mathbf{p}$.

It then follows that the result of these actions is the desired rotation about p. Algebraically, we have

$$\begin{aligned} f(\mathbf{x}) &= T_2(M(T_1\mathbf{x})) \\ &= T_2(M(\mathbf{x}-\mathbf{p})) \\ &= M(\mathbf{x}-\mathbf{p}) + \mathbf{p}. \end{aligned}$$

Example 3. Find the affine function that rotates the plane 90° counterclockwise about the point (1,1). Sketch its effect on the pentagon of Example 2.

Solution. With $M = \begin{bmatrix} 0 & -1 \\ 1 & 0 \end{bmatrix}$ and $\mathbf{p} = \begin{bmatrix} 1 \\ 1 \end{bmatrix}$ we have $f(\mathbf{x}) = M(\mathbf{x}-\mathbf{p}) + \mathbf{p}$ is the desired function. To be explicit, we write

$$\begin{aligned} f\begin{bmatrix} x \\ y \end{bmatrix} &= \begin{bmatrix} 0 & -1 \\ 1 & 0 \end{bmatrix}\begin{bmatrix} x-1 \\ y-1 \end{bmatrix} + \begin{bmatrix} 1 \\ 1 \end{bmatrix} \\ &= \begin{bmatrix} -y+1 \\ x-1 \end{bmatrix} + \begin{bmatrix} 1 \\ 1 \end{bmatrix} \\ &= \begin{bmatrix} 2-y \\ x \end{bmatrix}. \end{aligned}$$

Now, f simply fixes the vertex (1,1) of the pentagon and rotates.

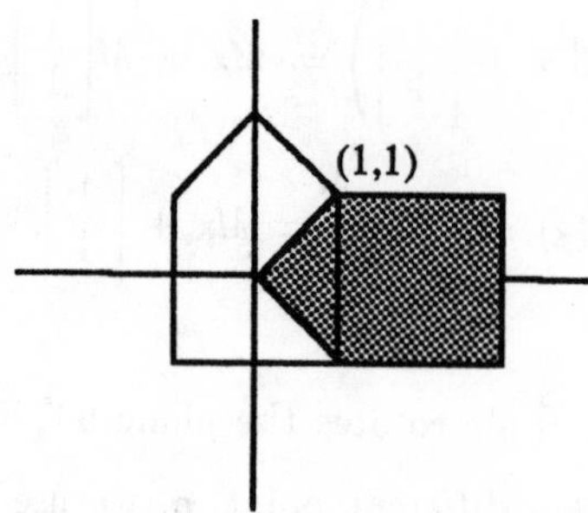

We will algebraically verify this by finding the image of the corners under f:

$$f\begin{bmatrix} 1 \\ 1 \end{bmatrix} = \begin{bmatrix} 1 \\ 1 \end{bmatrix}, \quad f\begin{bmatrix} 0 \\ 2 \end{bmatrix} = \begin{bmatrix} 0 \\ 0 \end{bmatrix}, \quad f\begin{bmatrix} -1 \\ 1 \end{bmatrix} = \begin{bmatrix} 1 \\ -1 \end{bmatrix}, \quad f\begin{bmatrix} -1 \\ -1 \end{bmatrix} = \begin{bmatrix} 3 \\ -1 \end{bmatrix}, \quad f\begin{bmatrix} 1 \\ -1 \end{bmatrix} = \begin{bmatrix} 3 \\ 1 \end{bmatrix}. \quad \square$$

It is sometimes necessary to construct affine functions that have prescribed mapping properties. For example in basic two dimensional computer graphics it is necessary to transform the "world" coordinate system to the fixed coordinate system of the computer screen. This can be done with a translation and a scaling. If we are working in the range $a \le x \le b$, $c \le y \le d$, and if our computer screen has lower left coordinates (n_1, m_1) and upper left coordinates (n_2, m_2), then we would like to map the rectangle on the left onto the rectangle on the right.

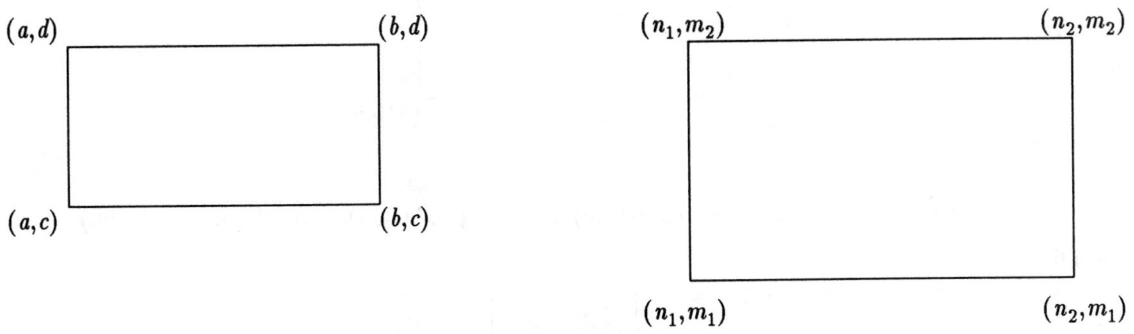

Since we usually want proportions to be preserved, we will use an affine function. In addition we will want (a,c) to go to (n_1,m_1) and (a,d) to go to (n_1,m_2).

Example 4. Find the affine function that takes the region $-1 \le x \le 1$, $0 \le y \le 2$ onto the rectangle

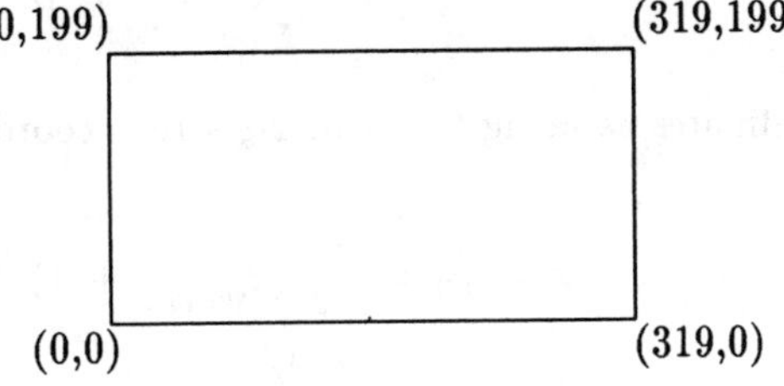

Solution. Our original rectangle is

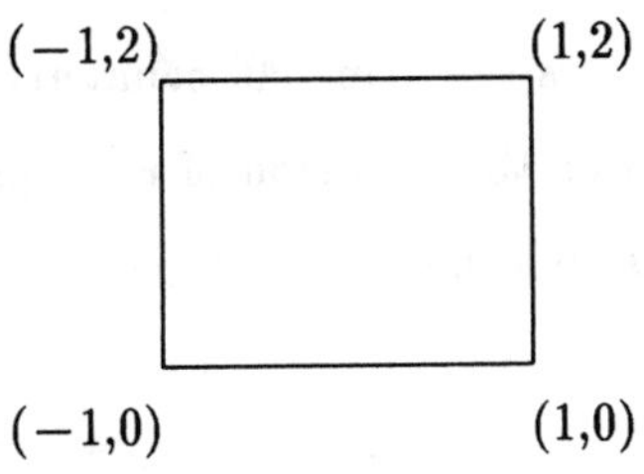

First, we move $(-1,0)$ to $(0,0)$ by the translation

$$T\mathbf{x} = \mathbf{x} - \begin{bmatrix} -1 \\ 0 \end{bmatrix}.$$

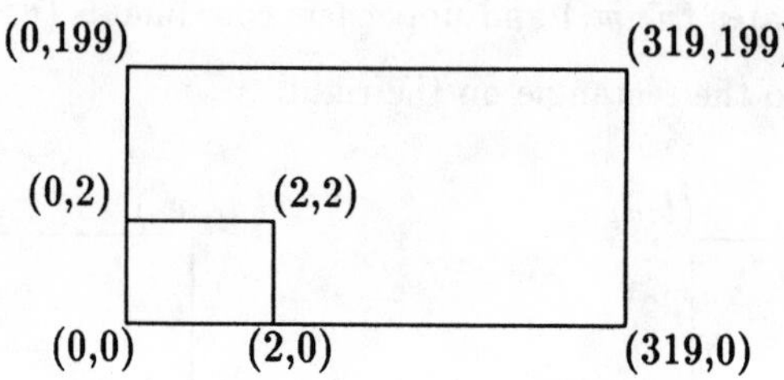

Then, we scale the x-axis so that (2,0) goes to (319,0) and the y-axis so that (0,2) goes to (0,199). This is accomplished by

$$D = \begin{bmatrix} \frac{319}{2} & 0 \\ 0 & \frac{199}{2} \end{bmatrix}.$$

So, the desired function is

$$f(\mathbf{x}) = D(T\mathbf{x}) = D\left(\mathbf{x} + \begin{bmatrix} 1 \\ 0 \end{bmatrix}\right).$$

If we think of x-y world coordinates as being taken to x-y screen coordinates, we can write

$$\begin{aligned} x_{screen} &= \frac{319}{2}(x_{world} + 1) \\ y_{screen} &= \frac{199}{2}(y_{world}). \end{aligned}$$ □

4.3. EXERCISES

1. Find the affine function that rotates the plane 90° counterclockwise about the point $\begin{bmatrix} 0 \\ 2 \end{bmatrix}$. Sketch the effect of this function on the pentagon of Example 1 and algebraically verify that the image of each corner is what it should be.

2. Find an affine function that takes the rectangle $-1 \le x \le 2$, $0 \le y \le 1$, onto the rectangle $0 \le u \le 10$, $0 \le v \le 5$.

3. Find an affine function that takes the rectangle of Example 4 onto the computer screen whose coordinate system is given by

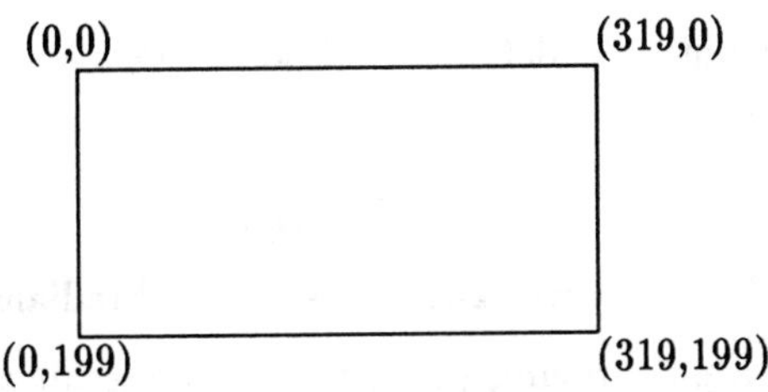

4. Find an affine function that takes the square with vertices $A = (0,0)$, $B = (1,0)$, $C = (1,1)$, $D = (0,1)$ onto the square with vertices $A' = (3,-2)$, $B' = (3,-1)$, $C' = (2,-1)$, $D' = (2,-2)$ with A going to A', B to B', etc. To what point does (7,5) go?

5. Find the image of the circle $x^2 + (y-1)^2 = 1$ under the affine function $f(\mathbf{x})$ of Example 4. Is the image a circle?

4.4. Rotations, Projections, and Reflections in $\mathbf{R}^2$

In order to manipulate two and three dimensional objects in computer graphical applications it is necessary to express various geometrical transformations algebraically. In this and the following section we discuss the linear functions defined by rigid motions and projections.

Rotations

We first consider the case of a counterclockwise rotation of θ radians about the origin. Observe that if you add two vectors and rotate their sum, you get the same result as rotating the vectors and then adding them.

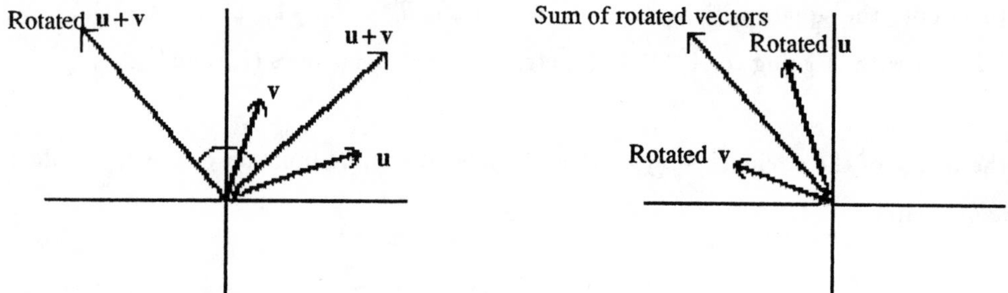

Figure 1.

Similarly, if you multiply a vector by a number and then rotate, you get the same result as rotating first and then multiplying. These observations show that the rotation is a linear function.

According to Theorem 2 of Section 4.2 there is a 2×2 matrix, call it R_θ, that gives this rotation. In addition the columns of R_θ are $R_\theta\mathbf{e}_1$ and $R_\theta\mathbf{e}_2$. Therefore, to know the algebraic rule for R_θ we need only know how $\mathbf{e}_1$ and $\mathbf{e}_2$ are rotated. This can be seen with elementary trigonometry.

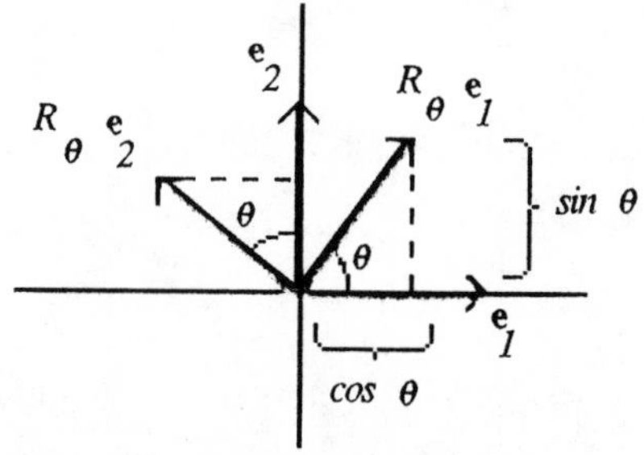

Figure 2.

The coordinates of $R_\theta \mathbf{e}_1$ are cos θ and sin θ and those of $R_\theta \mathbf{e}_2$ are $-$sin θ and cos θ.

Therefore,

$$R_\theta = \begin{bmatrix} \cos\theta & -\sin\theta \\ \sin\theta & \cos\theta \end{bmatrix}.$$

This generalizes the result of Section 4.2 that $\begin{bmatrix} 0 & -1 \\ 1 & 1 \end{bmatrix}$ rotates the plane by 90° in a counterclockwise manner.

Example 1. Find the matrix that rotates the plane counterclockwise about the origin by the $\frac{\pi}{6}$ radians.

Solution.

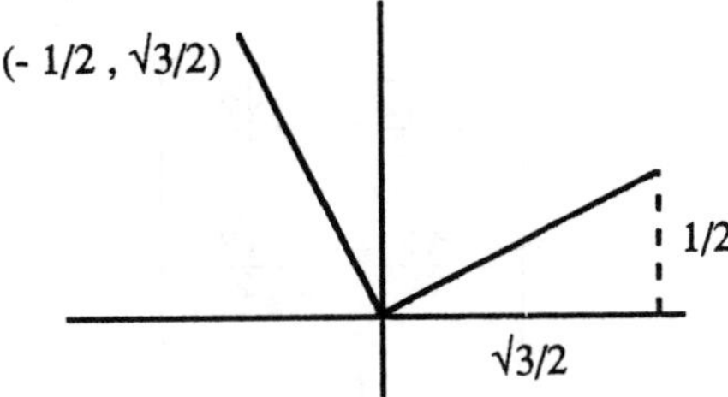

The first column of R_θ is

$$\begin{bmatrix} \frac{\sqrt{3}}{2} \\ \frac{1}{2} \end{bmatrix} \text{ and the second is } \begin{bmatrix} -\frac{1}{2} \\ \frac{\sqrt{3}}{2} \end{bmatrix}.$$

So

$$R_\theta = \frac{1}{2}\begin{bmatrix} \sqrt{3} & -1 \\ 1 & \sqrt{3} \end{bmatrix}.$$ □

Remarks. 1. The determinant of a rotation matrix is one. From the fact that rotations preserve distances, we see that they must preserve areas. So, the absolute value of the determinant must be one. (This requires no computation.) If we compute, we see *det* $R_\theta = \cos^2\theta + \sin^2\theta = 1$ for all θ.

2. Clockwise rotations are obtained by taking $\theta < 0$. So, $R_{-\frac{\pi}{2}}$ is the clockwise rotation by 90°. □

Example 2. Verify that the effect of a rotation of $\frac{\pi}{4}$ followed by a rotation of $\frac{\pi}{2}$ is the same as the effect of a rotation by $\frac{3\pi}{4}$.

Solution. In view of Theorem 3 of Section 4.2 we are to show that

$$R_{\frac{\pi}{2}} R_{\frac{\pi}{4}} = R_{\frac{3\pi}{4}}.$$

Here it is

$$\begin{bmatrix} 0 & -1 \\ 1 & 0 \end{bmatrix} \begin{bmatrix} \frac{\sqrt{2}}{2} & -\frac{\sqrt{2}}{2} \\ \frac{\sqrt{2}}{2} & \frac{\sqrt{2}}{2} \end{bmatrix} =$$

$$\begin{bmatrix} -\frac{\sqrt{2}}{2} & -\frac{\sqrt{2}}{2} \\ \frac{\sqrt{2}}{2} & -\frac{\sqrt{2}}{2} \end{bmatrix}$$

□

To rotate about a point p we proceed as in Section 4.2: first translate so that p goes to 0, next rotate, finally translate 0 to p.

Theorem 1. The counterclockwise rotation of $\mathbb{R}^2$ about the point p by θ radians is given by the function

$$f(\mathbf{x}) = R_\theta(\mathbf{x}-\mathbf{p}) + \mathbf{p}.$$

Example 3. Find an affine rotation that takes the pentagon with vertices $(\pm1, \pm1)$, $(0,2)$ to the figure

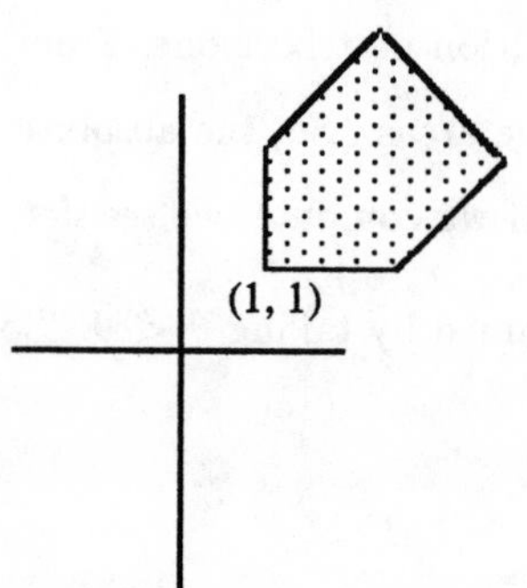

Solution. We start with

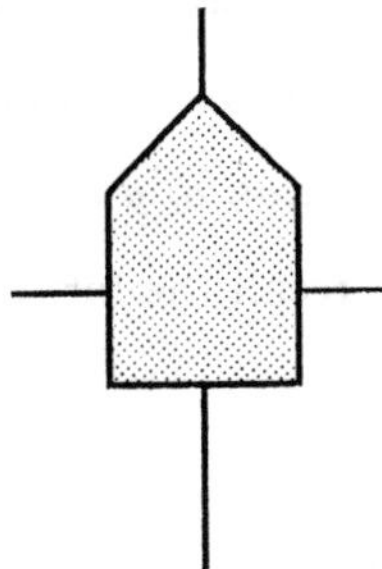

and first translate so that the tip (0,2) goes to (1,1). This is done by $T_1(\mathbf{x}) = \mathbf{x} + \begin{bmatrix} 1 \\ -1 \end{bmatrix}$ and gives

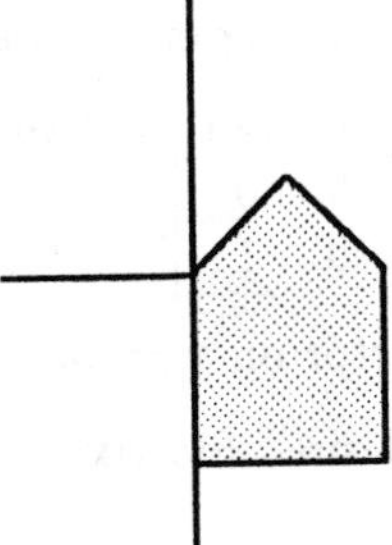

Now, we want to rotate about $\begin{bmatrix} 1 \\ 1 \end{bmatrix}$ by $\frac{3\pi}{4}$ radians in a counterclockwise manner. This is effected by

$$f(\mathbf{x}) = R_{\frac{3\pi}{4}}\left(\mathbf{x} - \begin{bmatrix} 1 \\ 1 \end{bmatrix}\right) + \begin{bmatrix} 1 \\ 1 \end{bmatrix}$$

and gives the desired configuration.

Our mapping is thus

$$\begin{aligned} g(\mathbf{x}) = f(T_1(\mathbf{x})) &= f\left(\mathbf{x} + \begin{bmatrix} 1 \\ -1 \end{bmatrix}\right) \\ &= R_{\frac{3\pi}{4}}\left(\mathbf{x} + \begin{bmatrix} -1 \\ 1 \end{bmatrix} - \begin{bmatrix} 1 \\ 1 \end{bmatrix}\right) + \begin{bmatrix} 1 \\ 1 \end{bmatrix} \\ &= R_{\frac{3\pi}{4}}\left(\mathbf{x} + \begin{bmatrix} -2 \\ 0 \end{bmatrix}\right) + \begin{bmatrix} 1 \\ 1 \end{bmatrix}. \end{aligned}$$

□

Projections onto Lines

Let $y = mx$ be the equation of a line in the Cartesian plane. Given a point p there is a unique point p^* on this line which is closest to p. p^* is called the *projection of* p *onto the line* $y = mx$. One can argue geometrically that this process is linear or one can recall the formula for projecting. If u is a *unit vector* on the line $y = mx$, then

$$\mathsf{p}^* = (\mathsf{u}^t\mathsf{p})\mathsf{u}.$$

Let's find the matrix that represents this action. First we consider two simple cases.

For the line $y = 0$ the projection of a point p is the point whose first coordinate is the same as p's and whose second coordinate is 0. The matrix that effects this change is $\begin{bmatrix} 1 & 0 \\ 0 & 0 \end{bmatrix}$. For the vertical line $\mathsf{x} = 0$ the associated projection matrix is $\begin{bmatrix} 0 & 0 \\ 0 & 1 \end{bmatrix}$.

In the general case we must first find a unit vector on $y = mx$. There are only two choices, we take $\mathsf{u} = \frac{1}{\sqrt{1+m^2}}\begin{bmatrix} 1 \\ m \end{bmatrix}$. Now, the first column of the desired matrix is

$$(\mathsf{u}^t\mathsf{e}_1)\ \mathsf{u} = \frac{1}{\sqrt{1+m^2}}\ \mathsf{u} = \frac{1}{1+m^2}\begin{bmatrix} 1 \\ m \end{bmatrix}.$$

The second column is

$$(\mathsf{u}^t\mathsf{e}_2)\ \mathsf{u} = \frac{m}{\sqrt{1+m^2}}\ \mathsf{u} = \frac{1}{1+m^2}\begin{bmatrix} m \\ m^2 \end{bmatrix}.$$

Theorem 2. The matrix that projects $\mathbb{R}^2$ onto the line $y = mx$ is

$$P_m = \begin{bmatrix} \frac{1}{1+m^2} & \frac{m}{1+m^2} \\ \frac{m}{1+m^2} & \frac{m^2}{1+m^2} \end{bmatrix}.$$

The mechanism for projecting onto lines in $\mathbb{R}^3$ is the same. A line in $\mathbb{R}^3$ which passes through the origin has the parametric form $\mathsf{x} = t\mathsf{d}$ where d is a vector giving the direction of the line. We can always arrange to have d be a unit vector. With this normalization the formula for the projection of a point $\mathbf{p}$ onto this line is

$$proj_{\mathbf{d}}\mathbf{p} = (\mathbf{p}^t\mathbf{d})\mathbf{d}.$$

Letting $\mathbf{p}$ be the standard basis vectors in $\mathbb{R}^3$ we see that the columns of the 3 by 3 matrix that represents the projection are from left to right $d_1\mathbf{d}$, $d_2\mathbf{d}$, and $d_3\mathbf{d}$ where the coordinates of $\mathbf{d}$ are taken to be d_1, d_2, and d_3. If we multiply this out, we get the matrix

$$\begin{bmatrix} d_1 d_1 & d_2 d_1 & d_3 d_1 \\ d_1 d_2 & d_2 d_2 & d_3 d_2 \\ d_1 d_3 & d_2 d_3 & d_3 d_3 \end{bmatrix}$$

which can be written in a somewhat simpler (factored) way as

$$\begin{bmatrix} d_1 \\ d_2 \\ d_3 \end{bmatrix} [d_1 \quad d_2 \quad d_3] = \mathbf{d}\mathbf{d}^t.$$

Theorem 3. If $\mathbf{d}$ is a unit vector, then the matrix M_d that projects $\mathbb{R}^3$ onto the line through the origin determined by $\mathbf{d}$ is

$$M_d = \mathbf{d}\mathbf{d}^t.$$

Reflections in $\mathbf{R}^2$

Definition. The *reflection* of a point $\mathbf{p}$ *through the line* L is the unique point $\mathbf{p}_R$ with the property that the line L is the perpendicular bisector of the line segment joining $\mathbf{p}$ and $\mathbf{p}_R$.

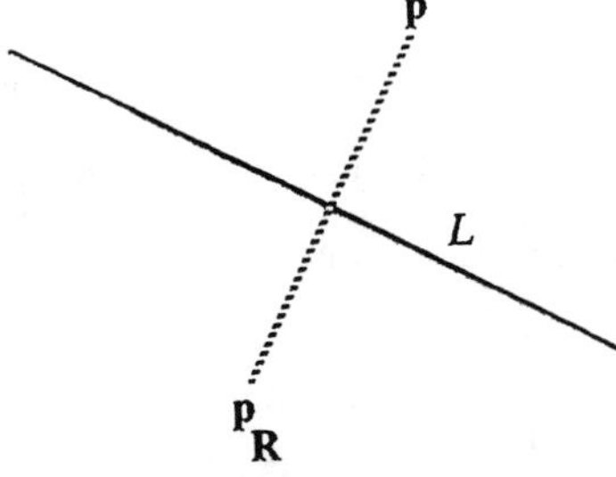

The process of reflecting through a fixed line preserves distances. If the line of reflection passes through the origin, then the origin is kept fixed and the reflection operator is linear. There are several ways to find the matrix that represents a reflection. We present one approach here and leave another for the exercises.

The key observation is that the midpoint of the line segment joining $\mathbf{p}$ and $\mathbf{p}_R$ must be the projection of $\mathbf{p}$ onto the line L.

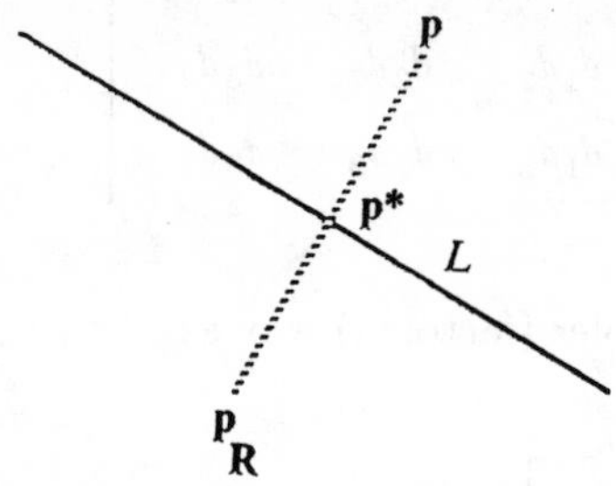

$$\mathbf{p}^* = (\mathbf{u}^t\mathbf{p})\mathbf{u}.$$

If we let P denote the matrix that projects onto L, then

$$\tfrac{1}{2}(\mathbf{p} + \mathbf{p}_R) = \mathbf{p}^* = P\mathbf{p}$$

or

$$\mathbf{p}_R = 2P\mathbf{p} - \mathbf{p} = (2P - I)\mathbf{p}.$$

Theorem 3. The matrix that represents reflection through the line $y = mx$ is $R_m = 2P_m - I$ where P_m is the projection onto the line.

Example 4. Find the matrix that represents reflection through the line $y = x$.

Solution. We can do this directly by inspection. Clearly $\mathbf{e}_1$ goes to $\mathbf{e}_2$ and $\mathbf{e}_2$ goes to $\mathbf{e}_1$. So the desired matrix is $\begin{bmatrix} 0 & 1 \\ 1 & 0 \end{bmatrix}$. Now, we will check this using Theorem 3. The projection matrix onto $y = x$ is

$$P_1 = \frac{1}{2}\begin{bmatrix} 1 & 1 \\ 1 & 1 \end{bmatrix}.$$

So, $$2P_1 - I = \begin{bmatrix} 1 & 1 \\ 1 & 1 \end{bmatrix} - \begin{bmatrix} 1 & 0 \\ 0 & 1 \end{bmatrix} = \begin{bmatrix} 0 & 1 \\ 1 & 0 \end{bmatrix}.$$ □

In numerical analysis reflection matrices are used to introduce zeroes into matrices.

Example 5. Find a reflection matrix that takes $\begin{bmatrix} 2 \\ 3 \end{bmatrix}$ into a vector of the form $\begin{bmatrix} \alpha \\ 0 \end{bmatrix}$.

Solution. Since rigid motions preserve lengths, we must have $2^2 + 3^2 = \alpha^2$.

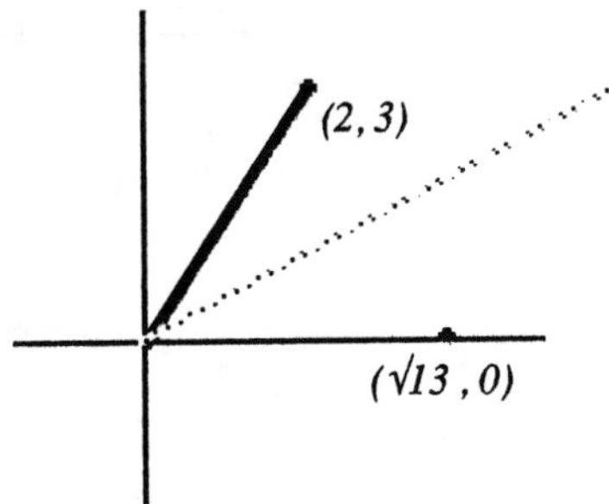

So, the desired reflection matrix takes $(\sqrt{13},0)$ to $(2,3)$ or $\mathbf{e}_1$ to $\left(\frac{2}{\sqrt{13}}, \frac{3}{\sqrt{13}}\right)$. This gives the first column. Now, reflections are rigid motions of the plane so they preserve angles. Therefore, $\mathbf{e}_2$ must go to a vector of length one perpendicular to $\left(\frac{2}{\sqrt{13}}, \frac{3}{\sqrt{13}}\right)$. So the desired matrix is

$$\text{either } \frac{1}{\sqrt{13}}\begin{bmatrix} 2 & -3 \\ 3 & 2 \end{bmatrix} \text{ or } \frac{1}{\sqrt{13}}\begin{bmatrix} 2 & 3 \\ 3 & -2 \end{bmatrix}.$$

The first candidate takes $\mathbf{e}_2$ to a point in the second quadrant so it can't be the correct choice. It is a rotation matrix in fact. So the desired matrix is

$$\frac{1}{\sqrt{13}}\begin{bmatrix} 2 & 3 \\ 3 & -2 \end{bmatrix}.$$ □

4.4. EXERCISES

1. Find an affine function that takes the pentagon with vertices $(\pm 1, \pm 1)$, $(0,2)$ to the given configurations. You should use only translations and rotations

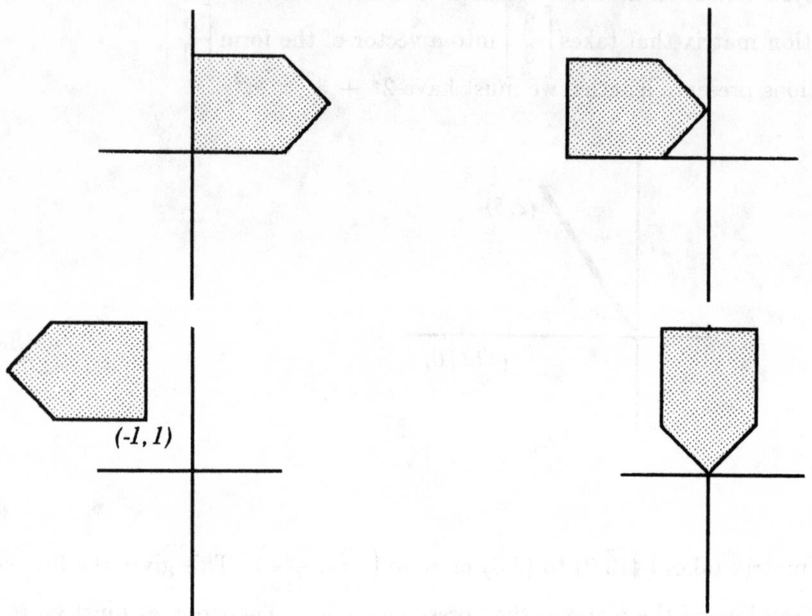

2. Is there a rotation matrix that transforms the square $ABCD$ to the square $A'B'C'D'$ with $A \to A'$, $B \to B'$, $C \to C'$, and $D \to D'$?

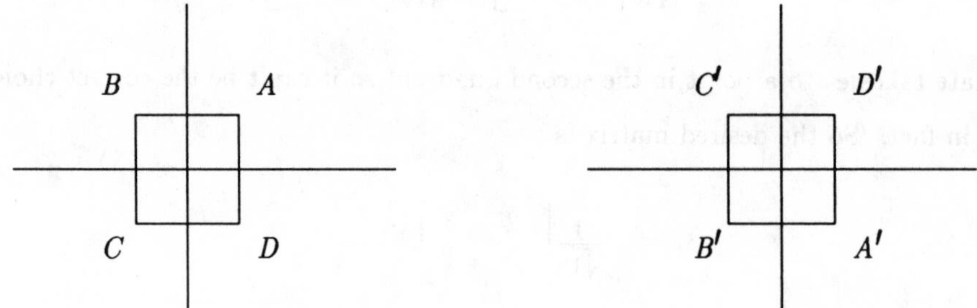

3. Find the matrix that projects $\mathbb{R}^3$ onto the line whose parametric equations are

(a) $x = t,\ y = 2t,\ z = -4t.$

(b) $x = 0,\ y = t,\ z = -t.$

4. The following properties of the projection matrix P_m appearing in Theorem 2 are evident from the geometric nature of P_m. Verify them algebraically.

(a) The column space of P_m is the line $y = mx$.

(b) The null space of P_m is the line $-my = x$.

(c) P_m does not have an inverse.

(d) $I - P_m$ is the projection onto the line $y = -\frac{1}{m}x$. That is, $I - P_m = P_{(-1/m)}$.

(e) $P_m P_m = P_m$.

5. The projection M_d appearing in Theorem 3 has properties similar to (a)-(d) in the previous exercise. Use your understanding of the geometry of projections to state what they should be.

6. The following properties of the reflection matrix R_m of Theorem 3 are easy to see geometrically. Verify them algebraically.

(a) $R_m R_m = I$, so $R_m^{-1} = R_m$.

(b) The column space of R_m is $\mathbb{R}^2$.

(d) $det\ R_m = -1$.

7. The reflection matrix for the line $y = mx$ can be represented in terms of sines and cosines. Let $m = \tan\,\theta$. Show that

$$R_m = \begin{bmatrix} \cos 2\theta & \sin 2\theta \\ \sin 2\theta & -\cos 2\theta \end{bmatrix}$$

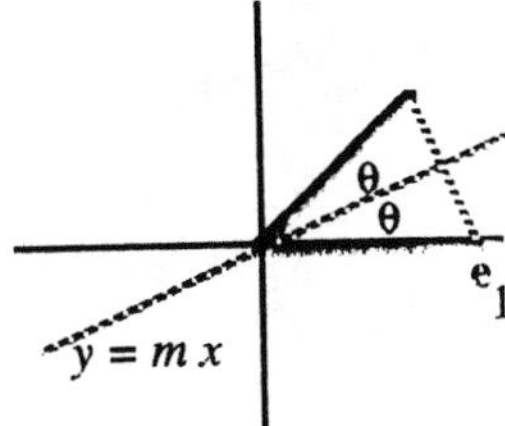

8. Exercise 7 can be used to show: Every rotation matrix is a product of two reflections. In fact, rotation by θ = (reflection through $y = mx$ where $m = \tan\frac{\theta}{2}$)(reflection through $y = 0$).

9. Find a reflection matrix that takes $\begin{bmatrix} 8 \\ 5 \end{bmatrix}$ to a multiple of $\begin{bmatrix} 1 \\ 0 \end{bmatrix}$.

10. Verify that the matrix $\frac{1}{\sqrt{13}}\begin{bmatrix} 2 & -3 \\ 3 & 2 \end{bmatrix}$ appearing in Example 5 is a rotation matrix.

4.5. Linear Functions in $\mathbb{R}^3$

In this section we will continue to study the representation of linear functions by matrices. The emphasis here is on functions that take $\mathbb{R}^3$ to itself.

Projections onto Planes

Let S be a plane in $\mathbb{R}^3$ containing the origin. For a given point $\mathbf{p}$ in $\mathbb{R}^3$ the projection of $\mathbf{p}$ onto S is the point on S which is closest to $\mathbf{p}$. We denote the projection of $\mathbf{p}$ onto S by $\mathbf{p}_S$. From geometry we know that $\mathbf{p}-\mathbf{p}_S$ must be perpendicular to S. This means that $\mathbf{p}-\mathbf{p}_S$ must be a multiple of a unit normal vector to S. By picking a unit normal so that $\mathbf{p}$ is a positive multiple of $\mathbf{n}$, we have the following.

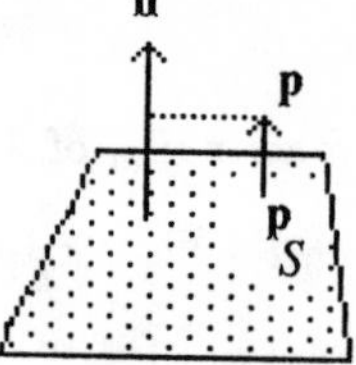

$\mathbf{p}-\mathbf{p}_S$ is the projection of $\mathbf{p}$ onto $\mathbf{n}$.

Since $\mathbf{p}-\mathbf{p}_S = proj_{\mathbf{n}}\mathbf{p}$, we can write

$$\mathbf{p}-\mathbf{p}_S = (\mathbf{n}^t\mathbf{p})\mathbf{n}$$

or

$$\mathbf{p}_S = \mathbf{p} - (\mathbf{n}^t\mathbf{p})\mathbf{n}.$$

In matrix terms, $\mathbf{p}$ is just $I\mathbf{p}$ and $(\mathbf{n}^t\mathbf{p})\mathbf{n}$ is the matrix $\mathbf{n}\mathbf{n}^t$ applied to $\mathbf{p}$.

Theorem 1. The matrix that represents the projection onto the plane through the origin whose unit normal vector is $\mathbf{n}$ is $I - \mathbf{n}\mathbf{n}^t$.

Example 1. Let S be the plane $x + y + z = 0$. We will find the matrix that represents the projection onto this plane.

The unit normal vector is $\left(\frac{1}{\sqrt{3}}, \frac{1}{\sqrt{3}}, \frac{1}{\sqrt{3}}\right)$ so $\mathbf{n}\mathbf{n}^t$ is

$$\begin{bmatrix} \frac{1}{3} & \frac{1}{3} & \frac{1}{3} \\ \frac{1}{3} & \frac{1}{3} & \frac{1}{3} \\ \frac{1}{3} & \frac{1}{3} & \frac{1}{3} \end{bmatrix}$$

and $I - \mathbf{n}\mathbf{n}^t$ is

$$\begin{bmatrix} \frac{2}{3} & -\frac{1}{3} & -\frac{1}{3} \\ -\frac{1}{3} & \frac{2}{3} & -\frac{1}{3} \\ -\frac{1}{3} & -\frac{1}{3} & \frac{2}{3} \end{bmatrix}.$$

At this time it is worth noting {the reader should verify this} that the range of the above matrix *is* the plane $x + y + z = 0$ and that its null space is the line through the origin determined by $\mathbf{n}$. □

Projections have many uses. Here is an example involving an overdetermined system of linear equations.

Example 2. Consider the overdetermined system

$$\begin{aligned} 3x + \quad y &= 4 \\ x - \quad 2y &= 0 \\ -4x + \quad y &= 2. \end{aligned}$$

The column space of the coefficient matrix is the plane $x + y + z = 0$ and the right hand side is not in the column space since $4 + 0 + 2 \neq 0$. Therefore, the above system has no solution. In this case, we can project the right hand side onto the column space and solve the resulting equations. This will give the *least squares solution* of the system. Now,

$$\begin{bmatrix} \frac{2}{3} & -\frac{1}{3} & -\frac{1}{3} \\ -\frac{1}{3} & \frac{2}{3} & -\frac{1}{3} \\ -\frac{1}{3} & -\frac{1}{3} & \frac{2}{3} \end{bmatrix} \begin{bmatrix} 4 \\ 0 \\ 2 \end{bmatrix} = \begin{bmatrix} 2 \\ -2 \\ 0 \end{bmatrix},$$

so we can solve the system

$$\begin{aligned} 3x + \quad y &= 2 \\ x - \quad 2y &= -2 \\ -4x + \quad y &= 0. \end{aligned}$$

to get $x = \frac{2}{7}$, $y = \frac{8}{7}$. □

Reflections Through Planes

The process of reflecting through a plane which passes through the origin is a linear process in $\mathbb{R}^3$ because it is a rigid motion that fixes the origin. The matrix that represents this function can be easily found using elementary geometry. If p is a point in $\mathbb{R}^3$ and S is a plane that contains the origin, then we denote the projection of p onto S by p_S. The reflection of p through S is the point, p_R, with the property that p_S is the midpoint of the line segment joining p and p_R. That is, $\mathsf{p} + \mathsf{p}_R = 2\mathsf{p}_S$. We could write $\mathsf{p}_R = 2\mathsf{p}_S - \mathsf{p}$ and obtain an expression for the reflection matrix in terms of the projection matrix for S. We will give another approach.

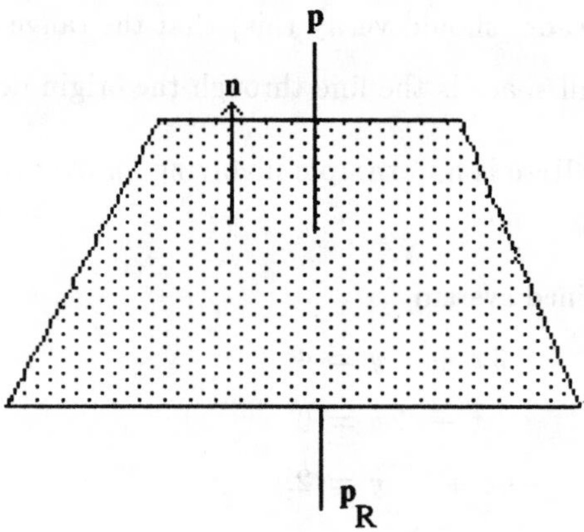

$\mathsf{p}-\mathsf{p}_R$ is a multiple of n.

Geometrically, we see that $\mathsf{p}-\mathsf{p}_R$ is a multiple of a unit normal to the plane, say n. That is,

$$\mathsf{p}-\mathsf{p}_R = \beta\mathsf{n}$$

for some number β where n has length 1 and is perpendicular to S. But we can choose n so that β must be the length of $\mathsf{p}-\mathsf{p}_R$ (this is just a matter of multiplying by -1 if necessary). It is clear from the definition of p_R that the distance of p from S is the same as that of p_R from S and that the length of $\mathsf{p}-\mathsf{p}_R$ is 2 times the distance of p from S. Now the distance of p from S is just $\mathsf{n}^t\mathsf{p}$. Hence $\beta = 2(\mathsf{n}^t\mathsf{p})$. This gives

$$\mathbf{p} - \mathbf{p}_R = 2(\mathbf{n}^t\mathbf{p})\mathbf{n}$$

or

$$\mathbf{p}_R = \mathbf{p} - 2(\mathbf{n}^t\mathbf{p})\mathbf{n}.$$

The expression $(\mathbf{n}^t\mathbf{p})\mathbf{n}$ can be recognized as the projection of $\mathbf{p}$ onto the line determined by $\mathbf{n}$. From earlier work we know that the matrix that represents this is $\mathbf{n}\mathbf{n}^t$.

Theorem 2. The matrix that represents the reflection through the plane whose unit normal vector is $\mathbf{n}$ is

$$I - 2\mathbf{n}\mathbf{n}^t.$$

Example 3. Find the matrix that represents reflection through the plane $x + y + z = 0$. The unit normal has all coordinates equal to $\frac{1}{\sqrt{3}}$ so as above $\mathbf{n}\mathbf{n}^t$ is

$$\begin{bmatrix} \frac{1}{3} & \frac{1}{3} & \frac{1}{3} \\ \frac{1}{3} & \frac{1}{3} & \frac{1}{3} \\ \frac{1}{3} & \frac{1}{3} & \frac{1}{3} \end{bmatrix}$$

and $I - 2\mathbf{n}\mathbf{n}^t$ is

$$\begin{bmatrix} \frac{1}{3} & -\frac{2}{3} & -\frac{2}{3} \\ -\frac{2}{3} & \frac{1}{3} & -\frac{2}{3} \\ -\frac{2}{3} & -\frac{2}{3} & \frac{1}{3} \end{bmatrix}.$$

Rotations that Fix a Coordinate Axis

We will find the matrix that keeps the x-axis fixed and rotates the y-z plane by θ radians counterclockwise as you look at it from the positive x-axis.

It is clear that if M is the matrix that does the prescribed rotation, then $M\mathbf{e}_1 = \mathbf{e}_1$ and the first column of M must be $\mathbf{e}_1$. The action of M on $\mathbf{e}_2$ can be discovered from the picture below.

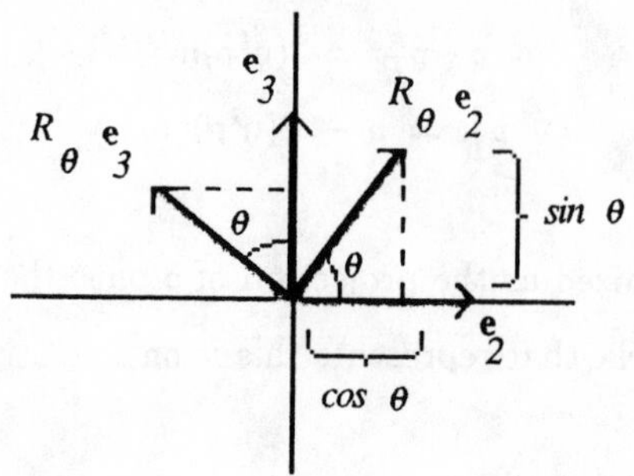

It follows that the second column of M is $\begin{bmatrix} 0 \\ \cos\theta \\ \sin\theta \end{bmatrix}$. Similarly the third column is $\begin{bmatrix} 0 \\ -\sin\theta \\ \cos\theta \end{bmatrix}$. Therefore, the desired matrix is

$$\begin{bmatrix} 0 & 0 & 0 \\ 0 & \cos\theta & -\sin\theta \\ 0 & \sin\theta & \cos\theta \end{bmatrix}$$

4.5. EXERCISES

1. Find the general form of a rotation that keeps the y-axis fixed and rotates the x-z plane by θ radians in a counterclockwise way when viewed from the positive y-axis.

2. Find the matrices that reflect through

   (a) the x-y plane (b) the x-z plane (c) the y-z plane.

   Find their pairwise products and interpret them geometrically.

3. Find the matrices that project onto the above planes.

4. Find the least squares solutions to the following systems using the approach of Example 2.

   (a) $$\begin{aligned} 3x + 4y &= 0 \\ -x + 5y &= 2 \\ 2x + 9y &= 4 \end{aligned}$$

   (b) $$\begin{aligned} x - 5y &= 2 \\ 3x + 3y &= -1 \\ -x - 13y &= 3. \end{aligned}$$

5. Let P be the matrix that projects $\mathbb{R}^3$ onto the plane $x + y + z = 0$. (a) Verify that the reflection through this plane is $2P - I$. (b) Verify that the projection onto the normal line is $I - P$.

CHAPTER 5 Eigenvalues and Eigenvectors

5.1. Geometric Introduction to Eigenvectors

One way to study the geometrical properties of linear functions is to consider how various simple subsets of the plane are moved. Let the square in $\mathbb{R}^2$ be determined by $\mathbf{e}_1$ and $\mathbf{e}_2$ be called the *unit square.*

Example 1. Let $A = \begin{bmatrix} 2 & 1 \\ 1 & 2 \end{bmatrix}$ and $D = \begin{bmatrix} 3 & 0 \\ 0 & 1 \end{bmatrix}$. The image of the unit square under A is the parallelogram with vertices (0,0), (2,1), (1,2), and (3,3). The image of the unit square under D is the *rectangle* with vertices (0,0), (3,0), (0,1), and (3,1).

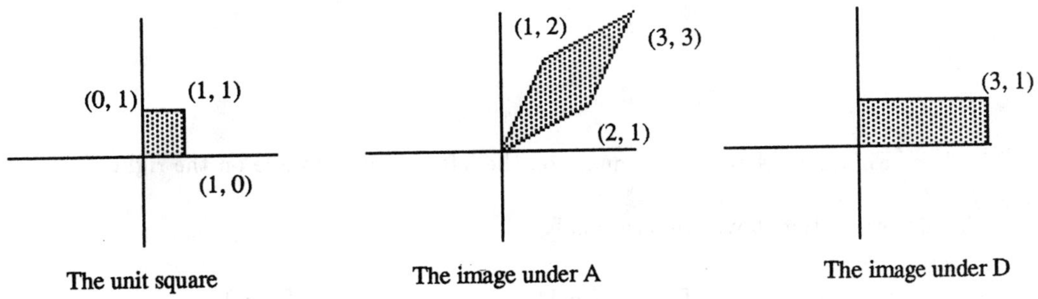

Figure 1. □

The geometric interpretation in this example did not say much more than the algebraic because all we did was multiply our matrices by the standard unit vectors $\mathbf{e}_1$ and $\mathbf{e}_2$, record our findings and then give them a geometric meaning. In spite of this fake attempt at geometry it cannot be denied that the geometry of the matrix D was simpler than that of A. In fact, D is simpler than A in other ways. For example, the system $D\mathbf{x} = \mathbf{b}$ is much easier to solve than $A\mathbf{x} = \mathbf{b}$. This is because there are no "connections" between the rows of D; the equations in a system having D as the coefficient matrix are *uncoupled* and thus require no Gaussian Elimination steps.

However, it is possible, by a special change of variables, to make the geometry of A as simple as that of D. An algebraic consequence will be the uncoupling of the rows of A.

We now make some observations about A. First, the vector $\begin{bmatrix} 1 \\ 1 \end{bmatrix}$ gets mapped to $\begin{bmatrix} 3 \\ 3 \end{bmatrix} = 3\begin{bmatrix} 1 \\ 1 \end{bmatrix}$. Second, the vector $\begin{bmatrix} -1 \\ 1 \end{bmatrix}$ gets mapped to itself. In other words the square with vertices

(0,0), (1,1), (−1,1), and (2,0) gets mapped to the rectangle with vertices (0,0), (3,3), (−1,1), and (2,4). Both of these have their sides on the lines $y = x$ and $y = -x$. Therefore, if we viewed these lines as our coordinate axes, the matrix A would act just like D.

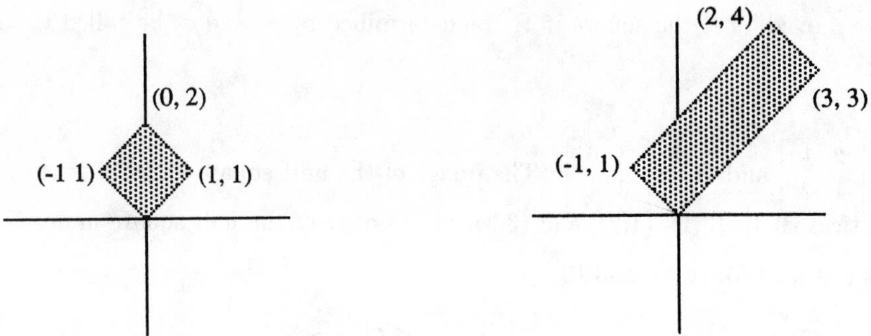

Figure 2. A takes the square on the left to the rectangle on the right.

The algebra of the above observations is

$$\begin{bmatrix} 2 & 1 \\ 1 & 2 \end{bmatrix}\begin{bmatrix} 1 \\ 1 \end{bmatrix} = \begin{bmatrix} 3 \\ 3 \end{bmatrix} \text{and} \begin{bmatrix} 2 & 1 \\ 1 & 2 \end{bmatrix}\begin{bmatrix} -1 \\ 1 \end{bmatrix} = \begin{bmatrix} -1 \\ 1 \end{bmatrix}.$$

These equations can be combined into one equation involving matrices:

$$\begin{bmatrix} 2 & 1 \\ 1 & 2 \end{bmatrix}\begin{bmatrix} 1 & -1 \\ 1 & 1 \end{bmatrix} = \begin{bmatrix} 3 & -1 \\ 3 & 1 \end{bmatrix}. \tag{1}$$

With $\mathbf{v}_1 = \begin{bmatrix} 1 \\ 1 \end{bmatrix}$ and $\mathbf{v}_2 = \begin{bmatrix} -1 \\ 1 \end{bmatrix}$ we have

$$A[\mathbf{v}_1 \quad \mathbf{v}_2] = [3\mathbf{v}_1 \quad \mathbf{v}_2] \tag{2}$$

or, equivalently,

$$A\mathbf{v}_1 = 3\mathbf{v}_1 \text{ and } A\mathbf{v}_2 = \mathbf{v}_2. \tag{3}$$

The equation (3) algebraically summarizes the special role that the vectors $\mathbf{v}_1$ and $\mathbf{v}_2$ play in describing the geometry of A.

Example 2. Let $B = \begin{bmatrix} -5 & 2 \\ 2 & -5 \end{bmatrix}$ and

$$\mathbf{u}_1 = \begin{bmatrix} \frac{1}{\sqrt{2}} \\ \frac{1}{\sqrt{2}} \end{bmatrix}, \quad \mathbf{u}_2 = \begin{bmatrix} -\frac{1}{\sqrt{2}} \\ \frac{1}{\sqrt{2}} \end{bmatrix}.$$

Verify that B takes the square determined by $\mathbf{u}_1$ and $\mathbf{u}_2$ to a rectangle whose sides are multiples of $\mathbf{u}_1$ and $\mathbf{u}_2$.

Solution. We compute $B\mathbf{u}_1$:

$$\begin{bmatrix} -5 & 2 \\ 2 & -5 \end{bmatrix} \begin{bmatrix} \frac{1}{\sqrt{2}} \\ \frac{1}{\sqrt{2}} \end{bmatrix} = \begin{bmatrix} -\frac{3}{\sqrt{2}} \\ -\frac{3}{\sqrt{2}} \end{bmatrix}$$

$$= -3 \begin{bmatrix} \frac{1}{\sqrt{2}} \\ \frac{1}{\sqrt{2}} \end{bmatrix}$$

and observe

$$B\mathbf{u}_1 = -3\mathbf{u}_1.$$

Similarly,

$$B\mathbf{u}_2 = -7\mathbf{u}_2.$$

So, B takes the square determined by $\mathbf{u}_1$ and $\mathbf{u}_2$ to the rectangle determined by $-3\mathbf{u}_1$ and $-7\mathbf{u}_2$:

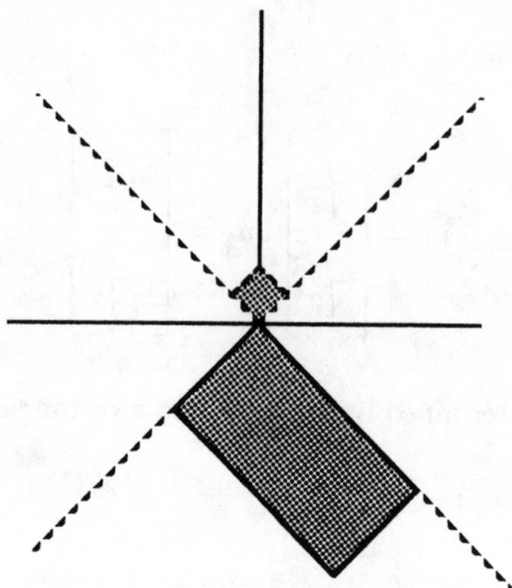

□

The matrices in the above examples had a special square on which they acted as a scaling. Relatively few matrices have such a square, but most 2×2 matrices have a distinguished parallelogram on which their geometric action is scaling.

Example 3. $M = \begin{bmatrix} 1 & 1 \\ -2 & 4 \end{bmatrix}$ and $\mathsf{u}_1 = \begin{bmatrix} 1 \\ 1 \end{bmatrix}$, $\mathsf{u}_2 = \begin{bmatrix} 1 \\ 2 \end{bmatrix}$. Verify that M takes the parallelogram determined by u_1 and u_2 onto a parallelogram whose sides are multiples of u_1 and u_2.

Solution. We compute $M\mathsf{u}_1$ and $M\mathsf{u}_2$:

$$\begin{bmatrix} 1 & 1 \\ -2 & 4 \end{bmatrix}\begin{bmatrix} 1 \\ 1 \end{bmatrix} = \begin{bmatrix} 2 \\ 2 \end{bmatrix} = 2\mathsf{u}_1 \quad \begin{bmatrix} 1 & 1 \\ -2 & 4 \end{bmatrix}\begin{bmatrix} 1 \\ 2 \end{bmatrix} = \begin{bmatrix} 3 \\ 6 \end{bmatrix} = 3\mathsf{u}_2.$$

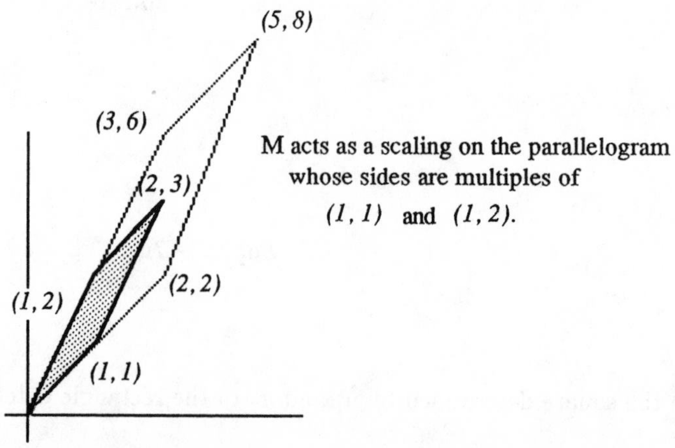
(5, 8)
(3, 6)
(2, 3)
(1, 2)
(2, 2)
(1, 1)
M acts as a scaling on the parallelogram
whose sides are multiples of
(1, 1) and (1, 2).

□

In the above examples, we were given directions that were left fixed by the matrices under consideration. In addition to the immediate geometric application of this knowledge there are important applications to problems which appear to have no direct contact with geometry. Before considering these applications we must introduce standard terminology.

Definition. A vector $\mathbf{v}$ is an *eigenvector* for $n \times n$ matrix A if $\mathbf{v}$ is ***not*** the zero vector and if there is a number λ so that $A\mathbf{v} = \lambda\mathbf{v}$. The number λ is called the *eigenvalue* for $\mathbf{v}$.

***Example* 4.** If D is an $n \times n$ diagonal matrix, then the diagonal entries of D are eigenvalues of D and the standard unit vectors are the eigenvectors. To see this observe $D\mathbf{e}_i = i^{th}$ *column of* $D = D_{ii}\mathbf{e}_i$. □

If A is a 2×2 matrix with eigenvalues λ_1 and λ_2 and respective eigenvectors $\mathbf{u}_1$ and $\mathbf{u}_2$, then the relations $A\mathbf{u}_1 = \lambda_1\mathbf{u}_1$ and $A\mathbf{u}_2 = \lambda_2\mathbf{u}$ can be written as

$$A[\mathbf{u}_1 \quad \mathbf{u}_2] = [\lambda_1\mathbf{u}_1 \quad \lambda_2\mathbf{u}_2]. \tag{4}$$

Letting

$$U = [\mathbf{u}_1 \quad \mathbf{u}_2] \text{ and } D = \begin{bmatrix} \lambda_1 & 0 \\ 0 & \lambda_2 \end{bmatrix}$$

we see that

$$[\lambda_1\mathbf{u}_1 \quad \lambda_2\mathbf{u}_2] = UD$$

so that (4) becomes

$$AU = UD. \tag{5}$$

If the columns of U are not multiples of each other — that is, if $\mathbf{u}_1$ and $\mathbf{u}_2$ are linearly independent, then U^{-1} exists and we may write

$$A = UDU^{-1}.$$

As a rule this formula is used symbolically and not computationally. In the general case we have

Theorem 1. If the $n \times n$ matrix A has n linearly independent eigenvectors, then there exists a matrix U and a diagonal matrix D so that $A = UDU^{-1}$. The columns of U are linearly independent eigenvectors of A and the diagonal entries of D are the eigenvalues of A.

Example 5. Verify equation (5) for the matrix of Example 3.

Solution. We have $M = \begin{bmatrix} 1 & 1 \\ -2 & 4 \end{bmatrix}$. The eigenvectors $\begin{bmatrix} 1 \\ 1 \end{bmatrix}$ and $\begin{bmatrix} 1 \\ 2 \end{bmatrix}$ with eigenvalues 2 and 3. So, we have $U = \begin{bmatrix} 1 & 1 \\ 1 & 2 \end{bmatrix}$ and $D = \begin{bmatrix} 2 & 0 \\ 0 & 3 \end{bmatrix}$. Now

$$MU = \begin{bmatrix} 1 & 1 \\ -2 & 4 \end{bmatrix} \begin{bmatrix} 1 & 1 \\ 1 & 2 \end{bmatrix} = \begin{bmatrix} 2 & 3 \\ 2 & 6 \end{bmatrix}$$

and

$$UD = \begin{bmatrix} 1 & 1 \\ 1 & 2 \end{bmatrix} \begin{bmatrix} 2 & 0 \\ 0 & 3 \end{bmatrix} = \begin{bmatrix} 2 & 3 \\ 2 & 6 \end{bmatrix}.$$ □

The zero vector is not an eigenvector. However, the number zero can be an eigenvalue. In fact zero is an eigenvalue of A if and only if the null space of A is not $\{\mathbf{0}\}$.

Example 6. Let P be the projection matrix $\begin{bmatrix} \frac{1}{2} & \frac{1}{2} \\ \frac{1}{2} & \frac{1}{2} \end{bmatrix}$. Show that 0 and 1 are eigenvalues of P and verify equation (5).

Solution. Members of the range of P are left fixed by P:

$$P\begin{bmatrix} 1 \\ 1 \end{bmatrix} = \begin{bmatrix} 1 \\ 1 \end{bmatrix}.$$

Thus, they are eigenvectors of P.

Members of the null space of P are sent to $\mathbf{0}$ by P so

$$P\begin{bmatrix} -1 \\ 1 \end{bmatrix} = \begin{bmatrix} 0 \\ 0 \end{bmatrix} = 0\begin{bmatrix} -1 \\ 1 \end{bmatrix}.$$

The matrices U and D in this case are

$$U = \begin{bmatrix} 1 & -1 \\ 1 & 1 \end{bmatrix} \quad D = \begin{bmatrix} 1 & 0 \\ 0 & 0 \end{bmatrix}.$$

We have

$$PU = \begin{bmatrix} \frac{1}{2} & \frac{1}{2} \\ \frac{1}{2} & \frac{1}{2} \end{bmatrix}\begin{bmatrix} 1 & -1 \\ 1 & 1 \end{bmatrix} = \begin{bmatrix} 1 & 0 \\ 1 & 0 \end{bmatrix}$$

and

$$UD = \begin{bmatrix} 1 & -1 \\ 1 & 1 \end{bmatrix}\begin{bmatrix} 1 & 0 \\ 0 & 0 \end{bmatrix} = \begin{bmatrix} 1 & 0 \\ 1 & 0 \end{bmatrix}.$$ □

Example 7. The 3×3 matrix $A = \begin{bmatrix} 2 & 1 & 1 \\ 1 & 2 & 1 \\ 1 & 1 & 2 \end{bmatrix}$ has eigenvalues 4 and 1 with $\begin{bmatrix} 1 \\ 1 \\ 1 \end{bmatrix}$ an eigenvector for 4 and $\begin{bmatrix} 1 \\ -1 \\ 0 \end{bmatrix}$ and $\begin{bmatrix} 1 \\ 1 \\ -2 \end{bmatrix}$ eigenvectors for 1. Find U and D so that Theorem 1 holds.

Solution. The columns of U are the eigenvectors and the diagonal entries of D are the corresponding eigenvalues. So

$$U = \begin{bmatrix} 1 & 1 & 1 \\ 1 & -1 & 1 \\ 1 & 0 & -2 \end{bmatrix} \text{ and } D = \begin{bmatrix} 4 & 0 & 0 \\ 0 & 1 & 0 \\ 0 & 0 & 1 \end{bmatrix}.$$

To check this we compute

$$AU = \begin{bmatrix} 2 & 1 & 1 \\ 1 & 2 & 1 \\ 1 & 1 & 2 \end{bmatrix}\begin{bmatrix} 1 & 1 & 1 \\ 1 & -1 & 1 \\ 1 & 0 & -2 \end{bmatrix} = \begin{bmatrix} 4 & 1 & 1 \\ 4 & -1 & 1 \\ 4 & 0 & -2 \end{bmatrix}$$

and

$$UD = \begin{bmatrix} 1 & 1 & 1 \\ 1 & -1 & 1 \\ 1 & 0 & -2 \end{bmatrix}\begin{bmatrix} 4 & 0 & 0 \\ 0 & 1 & 0 \\ 0 & 0 & 1 \end{bmatrix} = \begin{bmatrix} 4 & 1 & 1 \\ 4 & -1 & 1 \\ 4 & 0 & -2 \end{bmatrix}. \qquad \square$$

The final examples of this section show how Theorem 1 can be used to investigate the limiting behavior of powers of matrices.

Example 8. The matrix $S = \begin{bmatrix} \frac{1}{2} & \frac{1}{4} \\ \frac{1}{2} & \frac{3}{4} \end{bmatrix}$ has eigenvectors $\begin{bmatrix} 1 \\ -1 \end{bmatrix}$ and $\begin{bmatrix} 1 \\ 2 \end{bmatrix}$ with eigenvalues $\frac{1}{4}$ and 1, respectively. Show that the infinite sequence of vectors defined by x_0 = arbitrary, $\mathsf{x}_1 = S\mathsf{x}_0$, $\mathsf{x}_2 = S\mathsf{x}_1$, ..., $\mathsf{x}_{n+1} = S\mathsf{x}_n$, ..., converges to a multiple of $\begin{bmatrix} 1 \\ 2 \end{bmatrix}$.

Solution. First observe that $\mathsf{x}_n = S \cdot S \cdots S\mathsf{x}_0$ where S has been applied n-times. Now, $S = UDU^{-1}$ where

$$U = \begin{bmatrix} 1 & 1 \\ -1 & 2 \end{bmatrix} \text{ and } D = \begin{bmatrix} \frac{1}{4} & 0 \\ 0 & 1 \end{bmatrix}.$$

So, $S^n = S \cdot S \cdots S = UDU^{-1}UDU^{-1} \cdots UDU^{-1}$. The $U^{-1}U$ terms cancel and give $S^n = UD^nU^{-1}$ where

$$D^n = \begin{bmatrix} (\frac{1}{4})^n & 0 \\ 0 & 1 \end{bmatrix}.$$

Clearly as n $\to \infty$, $D^n \to \begin{bmatrix} 0 & 0 \\ 0 & 1 \end{bmatrix} = D^\infty$. Thus, as $n \to \infty$, x_n goes to $UD^\infty U^{-1}\mathsf{x}_0$. Since $U^{-1} = \frac{1}{3}\begin{bmatrix} 2 & -1 \\ 1 & 1 \end{bmatrix}$, we have

$$\lim_{n \to \infty} \mathsf{x}_n = \frac{1}{3}\begin{bmatrix} 1 & 1 \\ -1 & 2 \end{bmatrix}\begin{bmatrix} 0 & 0 \\ 0 & 1 \end{bmatrix}\begin{bmatrix} 2 & -1 \\ 1 & 1 \end{bmatrix}\begin{bmatrix} a \\ b \end{bmatrix}$$

$$= \frac{1}{3}\begin{bmatrix} 0 & 1 \\ 0 & 2 \end{bmatrix}\begin{bmatrix} 2a-b \\ a+b \end{bmatrix} = \frac{1}{3}\begin{bmatrix} a+b \\ 2a+2b \end{bmatrix} = \frac{a+b}{3}\begin{bmatrix} 1 \\ 2 \end{bmatrix}. \qquad \square$$

Example 9. Let M be the matrix of Example 3. Find M^5. Find formulas for the entries of M^n for integral n.

Solution. From Theorem 1 we have $M = UDU^{-1}$ where (cf. Example 5)

$$U = \begin{bmatrix} 1 & 1 \\ 1 & 2 \end{bmatrix} \text{ and } D = \begin{bmatrix} 2 & 0 \\ 0 & 3 \end{bmatrix}.$$

In general $M^n = UD^nU^{-1}$. (This holds for both positive and negative values of n in particular $M^{-1} = UD^{-1}U^{-1}$.) So, using

$$U^{-1} = \begin{bmatrix} 2 & -1 \\ -1 & 1 \end{bmatrix}$$

We have

$$M^n = \begin{bmatrix} 1 & 1 \\ 1 & 2 \end{bmatrix}\begin{bmatrix} 2^n & 0 \\ 0 & 3^n \end{bmatrix}\begin{bmatrix} 2 & -1 \\ -1 & 1 \end{bmatrix}$$

$$= \begin{bmatrix} 1 & 1 \\ 1 & 2 \end{bmatrix}\begin{bmatrix} 2{\cdot}2^n & -1{\cdot}2^n \\ -1{\cdot}3^n & 1{\cdot}3^n \end{bmatrix} = \begin{bmatrix} 2^{n+1}-3^n & 3^n-2^n \\ 2^{n+1}-2{\cdot}3^n & 2{\cdot}3^n-2^n \end{bmatrix}.$$

In particular

$$M^5 = \begin{bmatrix} 2^6-3^5 & 3^5-2^5 \\ 2^6-2{\cdot}3^5 & 2{\cdot}3^5-2^5 \end{bmatrix} = \begin{bmatrix} -179 & 211 \\ -422 & 454 \end{bmatrix}.$$ □

5.1. EXERCISES

1. What are the eigenvalues and eigenvectors of a 2×2 reflection matrix?

2. Use the method of Examples 8 and 9 to find formulas for the entries of A^n where A is as in Example 1. Verify that for $n = -1$ your formula gives A^{-1}.

3. Let B be as in Example 2. Find formulas for the entries of B^n. Use these formulas to find B^2 and B^{-2} and verify that $B^{-2}B^2 = I$.

4. Let A, D, and U be as in Example 7. Verify that the inverse of A is $UD^{-1}U^{-1}$.

5. Find the 2×2 matrix A that satisfies $A\mathbf{v} = 2\mathbf{v}$ and $A\mathbf{u} = \mathbf{u}$ where $\mathbf{v} = \begin{bmatrix} 0 \\ 1 \end{bmatrix}$ and $\mathbf{u} = \begin{bmatrix} 1 \\ 1 \end{bmatrix}$.

6. Find the 2×2 matrix B that satisfies $B\mathbf{x} = -\mathbf{x}$ and $B\mathbf{y} = \mathbf{0}$ where $\mathbf{x} = \begin{bmatrix} 2 \\ 3 \end{bmatrix}$ and $\mathbf{y} = \begin{bmatrix} -2 \\ 1 \end{bmatrix}$.

7. The column space and null space of matrix B of Problem 6 can be identified without any computations. What are they?

8. Find formulas for the entries of B^n where B is as in Problem 6.

9. Show that if $\mathbf{v}$ is an eigenvector for a matrix C, then so is any nonzero multiple of $\mathbf{v}$.

10. If a 2×2 matrix has *distinct* eigenvalues λ_1 and λ_2, then show that the corresponding eigenvectors cannot be multiples of each other.

11. If you attempt to solve the constrained extremal problem

$$\textit{minimize } ax^2 + 2bxy + cy^2 \textit{ subject to } x^2 + y^2 = 1$$

with the Lagrange Multiplier method, you will have to find the eigenvalues of a certain 2×2 matrix. What is that matrix? What is the relation between the eigenvalues of the matrix and the solution of the original extremal problem?

5.2. Computation of Eigenvalues and Eigenvectors

For 2×2 and 3×3 matrices eigenvalue-eigenvector computations can be efficiently carried out using a method based on determinants. For large matrices, other methods, which will not be discussed here, have been developed. However, the theoretical results which accompany the determinantal formulation are important for matrices of all sizes.

To compute the eigenvalues and eigenvectors of a given $n \times n$ matrix, we must solve the system $A\mathbf{x} - \lambda\mathbf{x} = \mathbf{0}$. The matrix form of this equation is

$$(A - \lambda I)\mathbf{x} = \mathbf{0}.$$

Since we want $\mathbf{x}$ not to be the zero vector {that's one of the rules}, we must choose λ so that there is more than one solution; that is, the null space of $A - \lambda I$ must have positive dimension. This will be the case if the column space of $A - \lambda I$ is not all of $\mathbb{R}^n$. This is equivalent to having the determinant of $A - \lambda I$ equal to zero.

Theorem 1. Let A be an $n \times n$ matrix. The number λ is an eigenvalue of A if and only if $det(A - \lambda I) = 0$. The corresponding eigenvector can be found by finding a nonzero solution of $(A - \lambda I)\mathbf{x} = \mathbf{0}$.

Definition. Let A be an $n \times n$ matrix. The *characteristic polynomial*, $p(\lambda)$, of A is the determinant of $A - \lambda I$: $p(\lambda) = det(A - \lambda I)$.

Theorem 2. The characteristic polynomial of an $n \times n$ matrix is a polynomial of degree n. The roots of the characteristic polynomial of A are the eigenvalues of A. In particular there are at most n distinct eigenvalues.

Example 1. Let $A = \begin{bmatrix} a & b \\ c & d \end{bmatrix}$ compute $det(A - \lambda I)$ and verify that it is a quadratic polynomial.

Solution. $A - \lambda I = \begin{bmatrix} a & b \\ c & d \end{bmatrix} - \lambda\begin{bmatrix} 1 & 0 \\ 0 & 1 \end{bmatrix} = \begin{bmatrix} a-\lambda & b \\ c & d-\lambda \end{bmatrix}$. So,

$$det(A - \lambda I) = (a-\lambda)(d-\lambda) - bc = \lambda^2 - (a+d)\lambda + ad - bc. \tag{1}$$

□

Remark. There are a number of points of interest even in the 2×2 case.

(1) Since quadratic polynomials can have complex roots, eigenvalues can be complex — even if the matrix has all real entries.

(2) If the roots of the quadratic (1) are λ_1, λ_2 we have

$$\lambda^2 - (a+d)\lambda + (ad-bc) = (\lambda-\lambda_1)(\lambda-\lambda_2) = \lambda^2 - (\lambda_1 + \lambda_2)\lambda + \lambda_1\lambda_2.$$

Therefore, $\lambda_1\lambda_2 = ad - bc = det\ A$ and $\lambda_1 + \lambda_2 = a + d$. That is, the product of the eigenvalues is the determinant and the sum of the eigenvalues is the sum of the diagonal entries. This is true in general.

Definition. The sum of the diagonal entries of an $n \times n$ matrix A is called the *trace* of A and is denoted by $tr(A)$.

Theorem 3. Let A be an $n \times n$ matrix. Then, $det\ A$ is the product of the eigenvalues of A and $tr(A)$ is the sum of the eigenvalues of A.

In Theorem 3 an eigenvalue that appears more than once as a root of the characteristic polynomial must be counted however many times it appears. See Example 3 below.

Example 2. Find the eigenvalues and eigenvectors of $A = \begin{bmatrix} 3 & 1 \\ 1 & 3 \end{bmatrix}$.

Solution.
$$A - \lambda I = \begin{bmatrix} 3 & 1 \\ 1 & 3 \end{bmatrix} - \begin{bmatrix} \lambda & 0 \\ 0 & \lambda \end{bmatrix} = \begin{bmatrix} 3-\lambda & 1 \\ 1 & 3-\lambda \end{bmatrix}$$

and $det(A - \lambda I) = (3-\lambda)(3-\lambda) - 1$. Setting $det(A - \lambda I) = 0$ and solving for λ gives $\lambda = 4$ and $\lambda = 2$. To find the eigenvalue for $\lambda = 4$ we must find a nonzero solution to

$$\begin{aligned} (3-4)x+y &= 0 \\ x+(3-4)y &= 0 \end{aligned}$$

This system just demands that $y = x$. So an eigenvector for the eigenvalue 4 is the vector $\begin{bmatrix} 1 \\ 1 \end{bmatrix}$ — or any nonzero multiple of it.

Similarly, to find an eigenvector for $\lambda = 2$ we solve

$$x + y = 0$$
$$x + y = 0.$$

This gives the relation $y = -x$ which in turn shows that $\begin{bmatrix} 1 \\ -1 \end{bmatrix}$ is an eigenvector for $\lambda = 2$.

We can summarize our findings by saying that $A = CDC^{-1}$ where

$$D = \begin{bmatrix} 4 & 0 \\ 0 & 2 \end{bmatrix} \text{ and } C = \begin{bmatrix} 1 & 1 \\ 1 & -1 \end{bmatrix}.$$

You can check this using the fact that $C^{-1} = \frac{1}{2}\begin{bmatrix} 1 & 1 \\ 1 & -1 \end{bmatrix}$.

Finally, we note that $tr(A) = 3 + 3 = 6 = 4 + 2 =$ *sum of eigenvalues* and $det\ A = 8 = 4{\cdot}2 =$ *product of eigenvalues.* □

Example 3. Let $B = \begin{bmatrix} 3 & 1 & 1 \\ 1 & 3 & 1 \\ 1 & 1 & 3 \end{bmatrix}$. Find the eigenvalues of B and as many linearly independent eigenvectors as possible.

Solution. $B - \lambda I = \begin{bmatrix} 3-\lambda & 1 & 1 \\ 1 & 3-\lambda & 1 \\ 1 & 1 & 3-\lambda \end{bmatrix}$ and

$$det(B - \lambda I) = (3 - \lambda)\{(3 - \lambda^2) - 1\} - (1)\{3 - \lambda - 1\} + (1)\{1 - (3 - \lambda)\}.$$

The equation

$$det(B - \lambda I) = 0$$

then simplifies to

$$(3 - \lambda)^3 - 3(3 - \lambda) + 2 = 0.$$

If we let $z = 3 - \lambda$, the equation takes the form

$$z^3 - 3z + 2 = 0.$$

Now, it is clear that $z = 1$ is a solution. It can also be seen that -2 is a root. To get the last root we should divide $z^3 - 3z + 1$ by the product $(z - 1)(z + 2) = z^2 + z - 2$. The quotient is $z-1$ which shows that 1 is a double root of the original equation.

Now, the eigenvalues of B are found by setting $\lambda = 3 - z$. This gives

$$\lambda = 5,\ 2,\ 2.$$

To get the eigenvector(s) for $\lambda = 2$, we look for nonzero solutions of the system $(B-3I)\mathbf{x} = \mathbf{0}$:

$$\begin{aligned} x + y + z &= 0 \\ x + y + z &= 0 \\ x + y + z &= 0. \end{aligned}$$

Therefore, the eigenvectors for $\lambda = 2$ are all of the nonzero vectors in the plane $x + y + z = 0$. For example, two independent such vectors are

$$\begin{bmatrix} 1 \\ 0 \\ -1 \end{bmatrix} \text{and} \begin{bmatrix} 1 \\ -2 \\ 1 \end{bmatrix}.$$

The fact that the number 2 occurred twice as a root of the characteristic polynomial was a hint that the set of eigenvectors might be a two dimensional set. {This is not always the case, however, see Exercise 1(b).}

For $\lambda = 5$ we have the system

$$\begin{aligned} -2x + y + z &= 0 \\ x - 2y + z &= 0 \\ x + y - 2z &= 0. \end{aligned}$$

We will use Gaussian Elimination to find the general solution. If we add $\frac{1}{2}$ times the first equation to the other two equations, we get

$$\begin{aligned} -2x + \quad y + \quad z &= 0 \\ -\tfrac{3}{2}\,y + \tfrac{3}{2}\,z &= 0 \\ \tfrac{3}{2}y - \tfrac{3}{2}\,z &= 0. \end{aligned}$$

From the last two equations we see that y *must equal* z. The first equation then reveals that x must be one-half of the sum of y and z; since $y = z$, we see that $x = z$ must follow. Therefore, the eigenvectors for $\lambda = 5$ are all nonzero vectors in which all three coordinates have the same value. For example, $\begin{bmatrix} 1 \\ 1 \\ 1 \end{bmatrix}$ is an eigenvector for $\lambda = 5$.

The algebraic results of this example can be summarized in the matrix equation

$$\begin{bmatrix} 3 & 1 & 1 \\ 1 & 3 & 1 \\ 1 & 1 & 3 \end{bmatrix} \begin{bmatrix} 1 & 1 & 1 \\ 0 & -2 & 1 \\ -1 & 1 & 1 \end{bmatrix} = \begin{bmatrix} 1 & 1 & 1 \\ 0 & -2 & 1 \\ -1 & 1 & 1 \end{bmatrix} \begin{bmatrix} 2 & 0 & 0 \\ 0 & 2 & 0 \\ 0 & 0 & 5 \end{bmatrix}.$$

The matrix that appears on both sides of this equation has the eigenvectors as its columns and the diagonal entries of the matrix on the extreme right are the corresponding eigenvalues.

Geometrically, we can say that the box that extends one unit in each of the eigenvector directions is sent to the box that extends 2 units in the directions of the first two eigenvectors and 5 units in the direction of the third eigenvector. The matrix multiplies the original volume by $(2)(2)(5) = 20$.

Also, note that $tr(B) = 3 + 3 + 3 = 2 + 2 + 5 =$ *sum of eigenvalues* and $det\ B = 20 =$ *product of eigenvalues.* □

5.2. EXERCISES

1. Find the eigenvalues and eigenvectors of the following matrices.

(a) $\begin{bmatrix} 3 & 1 \\ 2 & 4 \end{bmatrix}$ (b) $\begin{bmatrix} 1 & 1 \\ 0 & 1 \end{bmatrix}$ (c) $\begin{bmatrix} 0 & -1 \\ 1 & 0 \end{bmatrix}$ {eigenvalues are complex here}

(d) $\begin{bmatrix} -3 & -2 \\ 3 & 4 \end{bmatrix}$ (e) $\begin{bmatrix} 1 & 1 \\ -1 & -1 \end{bmatrix}$ (f) $\begin{bmatrix} 3 & -2 \\ 4 & -3 \end{bmatrix}$

(g) $\begin{bmatrix} 1 & 1 \\ 1 & 0 \end{bmatrix}$ (h) $\begin{bmatrix} 1 & 2 \\ 1 & 0 \end{bmatrix}$ (i) $\begin{bmatrix} 5 & 4 \\ 1 & 0 \end{bmatrix}$

2. Find the eigenvalues and eigenvectors

(a) $\begin{bmatrix} 0 & 1 & 0 \\ 0 & 0 & 1 \\ 1 & 0 & 0 \end{bmatrix}$ (b) $\begin{bmatrix} 1 & 1 & 1 \\ 1 & 1 & 1 \\ 1 & 1 & 1 \end{bmatrix}$ (c) $\begin{bmatrix} 3 & 2 & 0 \\ 2 & 0 & 0 \\ 1 & 0 & 2 \end{bmatrix}$.

3. Find matrices C and D so that $AC = CD$ for each matrix A in Exercise 1 for which this is possible. Discuss how the powers A^k behave as k goes to infinity.

4. Find a matrix B so that $B^2 = \begin{bmatrix} 4 & -1 \\ -1 & 4 \end{bmatrix}$. How many such B's are there?

5. If A is an $n \times n$ matrix and $AC = CD$ holds for some diagonal matrix D, then the columns of C must all be eigenvectors of A. Explain this.

6. Must a 2×2 matrix have two eigenvectors that determine a positive area? Explain. (Hint: Look at Problem 1(b).)

7. Suppose B is a 2×2 matrix for which $B^2 = 0$. Show that 0 must be an eigenvalue of B.

5.3. Quadratic Forms and the Second Derivative Test

The general form of a quadratic polynomial in two variables is

$$q(x,y) = ax^2 + 2bxy + cy^2 + ex + fy + g. \tag{1}$$

In three variables a quadratic polynomial takes the form

$$q(x,y,z) = ax^2 + 2bxy + 2cxz + dy^2 + 2eyz + fz^2 + gx + hy + iz + j. \tag{2}$$

If $\mathbf{x}$ is the column vector whose coordinates are x and y, then (1) can be written as

$$q(x,y) = \mathbf{x}^t A\mathbf{x} + ex + fy + g \tag{3}$$

where $A = \begin{bmatrix} a & b \\ b & c \end{bmatrix}$.

Similarly, with $\mathbf{x}^t = [x \;\; y \;\; z]$ we can write (2) as

$$q(x,y,z) = \mathbf{x}^t A\mathbf{x} + gx + hy + iz + j \tag{4}$$

where $A = \begin{bmatrix} a & b & c \\ b & d & e \\ c & e & f \end{bmatrix}$.

Notice that the both matrices above satisfy $A^t = A$. Such a matrix is called a *symmetric matrix.* The second derivative matrices of functions of several variables form an important class of symmetric matrices .

Definition. Let f be a real-valued function of n-variables. The *second derivative matrix* or *Hessian matrix* of f is the $n \times n$ matrix whose i-j^{th} entry is the function $\dfrac{\partial^2 f}{\partial x_i \partial x_j}$.

We denote the second derivative matrix by f''. If all functions in the matrix are evaluated at the point $\mathbf{p}$, the resulting matrix of numbers will be denoted by $f''(\mathbf{p})$. If f is sufficiently differentiable, $\dfrac{\partial^2 f}{\partial x_i \partial x_j} = \dfrac{\partial^2 f}{\partial x_j \partial x_i}$ so f'' will be symmetric.

Example 1. Let $f(x,y) = x^3 + x^2y + xy^3$. Then, $f_x = 3x^2 + 2xy + y^3$, $f_y = x^2 + 3xy^2$,

$f_{xx} = 6x + 2y$, $f_{yy} = 6xy$, $f_{xy} = f_{yx} = 2x + 3y^2$, so,

$$f'' = \begin{bmatrix} 6x+2y & 2x+3y^2 \\ 2x+3y^2 & 6xy \end{bmatrix}.$$

In particular,

$$f''(1,2) = \begin{bmatrix} 10 & 14 \\ 14 & 12 \end{bmatrix}. \qquad \square$$

The Hessian matrix can be used to find quadratic approximations to f.

Definition. If f is a real-valued function of n-variables, then *the best quadratic approximation to f at the point* $\mathbf{p}$ is the function

$$q(\mathbf{x}) = f(\mathbf{p}) + (\nabla f(\mathbf{p}))^t(\mathbf{x}-\mathbf{p}) + \tfrac{1}{2}(\mathbf{x}-\mathbf{p})^t f''(\mathbf{p})(\mathbf{x}-\mathbf{p}).$$

The best quadratic function and its first and second order partials agree with the values of f and its corresponding partials at the points $\mathbf{p}$. This is the content of

Theorem 1. If q is the best quadratic approximation to f at the point $\mathbf{p}$ and if f is "smooth enough," then

(a) $\quad q(\mathbf{p}) = f(\mathbf{p})$, $\nabla q(\mathbf{p}) = \nabla f(\mathbf{p})$, and $q''(\mathbf{p}) = f''(\mathbf{p})$

and (b) $\quad$ as $\mathbf{x}$ goes to $\mathbf{p}$, $|f(\mathbf{x}) - q(\mathbf{x})|$ goes to zero faster than $|\mathbf{x}-\mathbf{p}|^2$ does.

Example 2. Find the best quadratic approximation to $f(x,y) = x^3 + x^2y + xy^3$ at $\mathbf{p} = (1,2)$.

Solution. The partials of f and $f''(1,2)$ were computed in Example 1. We have

$$\nabla f(1,2) = \begin{bmatrix} f_x(1,2) \\ f_y(1,2) \end{bmatrix} = \begin{bmatrix} 15 \\ 13 \end{bmatrix}, \quad \mathbf{x}-\mathbf{p} = \begin{bmatrix} x-1 \\ y-2 \end{bmatrix}, \quad f(1,2) = 11.$$

So,

$$q(x,y) = 11 + [15 \quad 13]\begin{bmatrix} x-1 \\ y-2 \end{bmatrix} + \tfrac{1}{2}[x-1 \quad y-2]\begin{bmatrix} 10 & 14 \\ 14 & 12 \end{bmatrix}\begin{bmatrix} x-1 \\ y-2 \end{bmatrix}$$

or $$q(x,y) = 11 + 15(x-1) + 13(y-2) + \tfrac{1}{2}10(x-1)^2 + 14(x-1)(y-2) + \tfrac{1}{2}2(y-2)$$

or $$q(x,y) = 11 + 15(x-1) + 13(y-2) + 5(x-1)^2 + 14(x-1)(y-2) + 6(y-2)^2.$$

In practice, one would ***not*** expand this expression for q. □

The best quadratic approximation to a function at a critical point can sometimes be used to determine the nature of the critical point. For if p is a critical point of f, then $\nabla f(\mathsf{p}) = \mathbf{0}$ and for x close to p we have the approximate equality

$$f(\mathsf{x}) \approx f(\mathsf{p}) + \tfrac{1}{2}(\mathsf{x}-\mathsf{p})^t f''(\mathsf{p})(\mathsf{x}-\mathsf{p}).$$

If the pure quadratic term, $(\mathsf{x}-\mathsf{p})^t f''(\mathsf{p})(\mathsf{x}-\mathsf{p})$, were greater than zero for all x except $\mathsf{x} = \mathsf{p}$, then we would have that for all x close to p $f(\mathsf{x}) \geq f(\mathsf{p})$; thus, f would have a local minimum at p. Similarly, if the quadratic term were less than zero (except for $\mathsf{x} = \mathsf{p}$), f would have a local maximum at p.

{If $(\mathsf{x}-\mathsf{p})^t f''(\mathsf{p})(\mathsf{x}-\mathsf{p})$ equals zero for some values of $\mathsf{x} \neq \mathsf{p}$, then we could not conclude that f has a local minimum at p. The reason for this is hidden in the "$\approx$" sign. It can be shown that

$$f(\mathsf{x}) = f(\mathsf{p}) + (\mathsf{x}-\mathsf{p})^t f''(\eta)(\mathsf{x}-\mathsf{p})$$

where η is some (unknown) point depending on x and p. If $(\mathsf{x}-\mathsf{p})^t f''(\mathsf{p})(\mathsf{x}-\mathsf{p})$ is positive for all values of x except $\mathsf{x} = \mathsf{p}$, then we can guarantee that $(\mathsf{x}-\mathsf{p})^t f''(\eta)(\mathsf{x}-\mathsf{p})$ is also positive provided η is close enough to p. However, if $(\mathsf{x}-\mathsf{p})^t f''(\mathsf{p})(\mathsf{x}-\mathsf{p})$ equals zero for some $\mathsf{x} \neq \mathsf{p}$, then we cannot guarantee anything about the sign behavior of $(\mathsf{x}-\mathsf{p})^t f''(\eta)(\mathsf{x}-\mathsf{p})$.}

Now a slight notational simplification is possible. We are interested in the sign of $(\mathsf{x}-\mathsf{p})^t f''(\mathsf{p})(\mathsf{x}-\mathsf{p})$ as x varies. We may think of x as being p plus something; that is, we can let $\mathsf{x} = \mathsf{p} + \mathsf{X}$ so that $\mathsf{x} - \mathsf{p} = \mathsf{X}$. Thus, we should study the sign pattern of $\mathsf{X}^t f''(\mathsf{p})\mathsf{X}$. But now we could just as well rename X to be x. We will do this and study functions of the form

$$g(\mathsf{x}) = \mathsf{x}^t A \mathsf{x}$$

where A is a symmetric matrix. It suffices to study the level sets of such functions. Essentially, if the level sets are bounded sets – like ellipses, then the function $g(\mathsf{x})$ is of one sign; if the level sets are unbounded sets, then the function $g(\mathsf{x})$ changes sign. We will concentrate on the two dimensional case.

Diagonalization of Quadratic Forms

We will use eigenvectors and eigenvalues to rewrite quadratic equations in two variables in a way that make their graphs readily recognizable. First, we recall that in the $u-v$ coordinate system an equation of the form

$$\lambda u^2 + \mu v^2 = 1 \tag{5}$$

represents an ellipse if λ and μ are both positive and a hyperbola if λ and μ have opposite signs. If λ and μ are both negative then the set represented by the equation is the empty set. If $\lambda\mu = 0$ and neither λ nor μ are zero, then the equation represents two parallel lines.

We are going to classify the curves given by equations of the form

$$ax^2 + 2bxy + cy^2 = 1 \tag{6}$$

by finding a coordinate system in which this curve has an equation of the form (5). The new coordinate system will be determined by the eigenvectors of an appropriate matrix. The eigenvalues of this matrix will determine the nature of the curve. The key observation made above is that the left hand side of equation (6) can be viewed as the result of computing $\mathsf{x}^t A \mathsf{x}$ for the matrix

$$A = \begin{bmatrix} a & b \\ b & c \end{bmatrix}.$$

A very special property of the eigenvectors of symmetric matrices makes eigenanalysis work.

Theorem 1. Let A be an $n \times n$ symmetric matrix. Then, A has n eigenvectors which are mutually perpendicular.

If these eigenvectors are normalized to have length 1 and $U = [\mathbf{u}_1 \quad \cdots \quad \mathbf{u}_n]$ is the matrix whose columns are the normalized eigenvectors, then $U^{-1} = U^t$.

Example 3. For $\begin{bmatrix} 2 & 1 \\ 1 & 2 \end{bmatrix}$ the vectors $\begin{bmatrix} \frac{1}{\sqrt{2}} \\ \frac{1}{\sqrt{2}} \end{bmatrix}, \begin{bmatrix} -\frac{1}{\sqrt{2}} \\ \frac{1}{\sqrt{2}} \end{bmatrix}$ are mutually perpendicular eigenvectors of length one. With

$$U = \begin{bmatrix} \frac{1}{\sqrt{2}} & -\frac{1}{\sqrt{2}} \\ \frac{1}{\sqrt{2}} & \frac{1}{\sqrt{2}} \end{bmatrix}$$

we have

$$U^{-1} = \begin{bmatrix} \frac{1}{\sqrt{2}} & \frac{1}{\sqrt{2}} \\ -\frac{1}{\sqrt{2}} & \frac{1}{\sqrt{2}} \end{bmatrix} = U^t.$$

It is also worth noting that U is a rotation by $\frac{\pi}{4}$. □

Theorem 2. Let A be an $n \times n$ symmetric matrix. Then, there is a diagonal matrix D and an invertible matrix U so that

$$A = UDU^t.$$

The next example illustrates how this helps determine the nature of quadratic curves.

Example 4. Let $A = \begin{bmatrix} 2 & -1 \\ -1 & 2 \end{bmatrix}$. Show that the curve $2x_1^2 - 2x_1x_2 + 2x_2^2 = \mathbf{x}^t A\mathbf{x} = 1$ is an ellipse and sketch it.

Solution. The eigenvalues of A are 3 and 1 with normalized eigenvectors $\mathbf{v}_1 = \begin{bmatrix} -\frac{1}{\sqrt{2}} \\ \frac{1}{\sqrt{2}} \end{bmatrix}$ and

$\mathbf{v}_2 = \begin{bmatrix} \frac{1}{\sqrt{2}} \\ \frac{1}{\sqrt{2}} \end{bmatrix}$. We write $A = UDU^t$ where

$$U = \begin{bmatrix} -\frac{1}{\sqrt{2}} & \frac{1}{\sqrt{2}} \\ \frac{1}{\sqrt{2}} & \frac{1}{\sqrt{2}} \end{bmatrix} \text{ and } D = \begin{bmatrix} 3 & 0 \\ 0 & 1 \end{bmatrix}.$$

Now, $\mathbf{x}^t A\mathbf{x} = \mathbf{x}^t UDU^t\mathbf{x} = (U^t\mathbf{x})^t D(U^t\mathbf{x})$. So, if we let $\mathbf{u} = U^t\mathbf{x}$ where $\mathbf{u} = \begin{bmatrix} u_1 \\ u_2 \end{bmatrix}$, we have

$$\mathbf{x}^t A\mathbf{x} = \mathbf{u}D\mathbf{u} = [u_1 \quad u_2]\begin{bmatrix} 3 & 0 \\ 0 & 1 \end{bmatrix}\begin{bmatrix} u_1 \\ u_2 \end{bmatrix} = 3u_1^2 + u_2^2.$$

Therefore, the $x - x_2$ equation

$$2x_1^2 - 2x_1x_2 + 2x_2^2 = 1 \tag{7}$$

is equivalent to

$$3u_1^2 + u_2^2 = 1, \tag{8}$$

which is an ellipse in the $u_1 - u_2$ coordinate system.

Now, how is the $u_1 - u_2$ coordinate system related to the $x_1 - x_2$ system? This can be answered by analyzing the relation $\mathbf{u} = U^t\mathbf{x}$. Since $U = [\mathbf{v}_1 \quad \mathbf{v}_2]$,

$$\mathbf{u} = \begin{bmatrix} u_1 \\ u_2 \end{bmatrix} = U^t\mathbf{x} = \begin{bmatrix} \mathbf{v}_1^t\mathbf{x} \\ \mathbf{v}_2^t\mathbf{x} \end{bmatrix}.$$

Observe:

$$u_1\text{-}axis = \{\mathbf{u}\colon u_2 = 0\} = \{\mathbf{x}\colon \mathbf{v}_2^t\mathbf{x} = 0\} = \{\mathbf{x}\colon x_1 + x_2 = 0\}$$

$$u_2\text{-}axis = \{\mathbf{u}\colon u_1 = 0\} = \{\mathbf{x}\colon \mathbf{v}_1^t\mathbf{x} = 0\} = \{\mathbf{x}\colon x_1 - x_2 = 0\}.$$

The last equality on each line uses the specific values of $\mathbf{v}_1$ and $\mathbf{v}_2$. So, we have

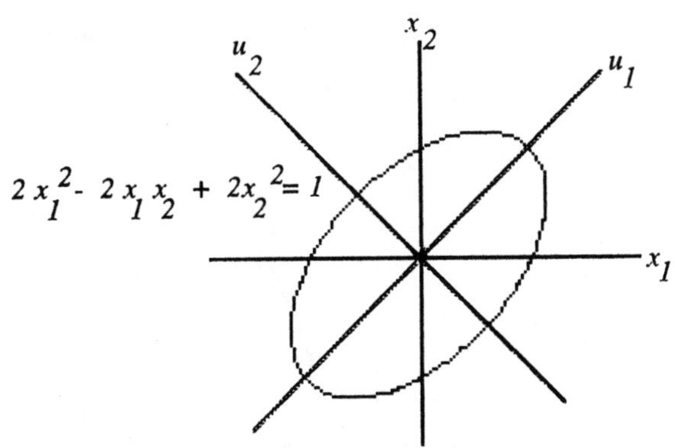

□

The reasoning used above gives

Theorem 3. The curve $ax_1^2 + 2bx_1x_2 + cx_2^2 = 1$ is identical to the curve

$$\lambda_1 u_1^2 + \lambda_2 u_2^2 = 1$$

where λ_1 and λ_2 are the eigenvalues of

$$\begin{bmatrix} a & b \\ b & c \end{bmatrix}$$

and the u_i-axis is the line determined by the eigenvector for λ_i, $i = 1,2$.

Example 5. Sketch the curve $x_1^2 + 4x_1x_2 - 2x_2^2 = 5$.

Solution. We consider the matrix $A = \begin{bmatrix} 1 & 2 \\ 2 & -2 \end{bmatrix}$. The eigenvalues of A are the solutions $(1-\lambda)(-2-\lambda) - 2 = 0$, i.e., $\lambda^2 + \lambda - 6 = 0$. So, the eigenvalues are 2 and -3. Thus, the curve is a hyperbola. An eigenvector for 2 is $\mathbf{v}_1 = \begin{bmatrix} s \\ t \end{bmatrix}$ where

$$-s + 2t = 0$$
$$2s - 4t = 0.$$

Therefore, $\mathbf{v}_1 = \begin{bmatrix} 2 \\ 1 \end{bmatrix}$. The eigenvectors for -3 are perpendicular to $\mathbf{v}_1$, so we may take $\mathbf{v}_2 = \begin{bmatrix} -1 \\ 2 \end{bmatrix}$. So, with the u_1-axis containing $\mathbf{v}_1$ and the u_2-axis containing $\mathbf{v}_2$ we have the original equation is identical to $2u_1^2 - 3u_2^2 = 5$.

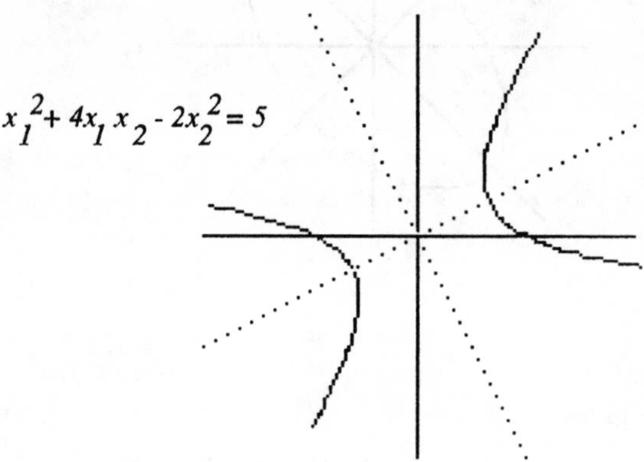

□

Remark. The above discussion carries over to higher dimensions. In three dimensions we consider the problem of putting equations of the form

$$ax_1^2 + bx_2^2 + cx_3^2 + 2dx_1x_2 + 2ex_1x_3 + 2fx_2x_3 = 1$$

into a standard form

$$d_1u_1^2 + du_2^2 + d_3u_3^2 = 1$$

by an appropriate change of variables. It turns out that the numbers d_1, d_2, and d_3 are the eigenvalues of the 3×3 symmetric matrix

$$\begin{bmatrix} a & d & e \\ d & b & f \\ e & f & c \end{bmatrix}$$

and the new coordinate axes are pointed out by the corresponding eigenvectors which are mutually perpendicular. One can then classify the surface by noting the sign pattern of the eigenvalues.

We finish this section by returning to the best quadratic approximation.

The Second Derivative Test

Let f be a real-valued function of n-variables with a critical point $\mathbf{p}$. For $\mathbf{x}$ close to $\mathbf{p}$ we have

$$f(\mathbf{x}) \approx f(\mathbf{p}) + \tfrac{1}{2}(\mathbf{x}-\mathbf{p})^t f''(\mathbf{p})(\mathbf{x}-\mathbf{p}).$$

If all the eigenvalues of $f''(\mathbf{p})$ are of one sign, then the level sets of $\mathbf{x}^t f''(\mathbf{p})\mathbf{x}$ are ellipsoids and f has a *local minimum* (if all eigenvalues are positive) or a *local maximum* (if all eigenvalues are negative). If there are eigenvalues of both signs, then the level sets are hyperboloids and f has a *saddle point* at $\mathbf{p}$.

Notice that in dimensions greater than two we can have some eigenvalues equal to zero and still be able to classify the critical point. Also note that the determinant plays no direct role in the higher dimensional case.

5.3. EXERCISES

1. Find the best quadratic approximation of the following at the indicated points

   (a) $f(x,y) = x^2 \sin y, \qquad (x,y) = (1, \frac{\pi}{2})$

   (b) $g(x,y,z) = x^3 y + y^2 z + xz, \qquad (x,y,z) = (0,0,0)$

   (c) $h(x,y) = x^4 + xy^3 + x^2y^2, \qquad (x,y) = (0,1)$

2. Sketch the curve $5x^2 - 4xy + 2y^2 = 1$.

3. Sketch the curves represented by $4x^2 + 2\lambda xy + 4y^2 = 1$ for the following values of λ: 0, 2, -2, 4, 5, -5.

4. For what values of λ is the curve $2x^2 + 2\lambda xy + 3y^2 = 1$ an ellipse? What are its axes of symmetry?

5. In Example 4 the relation $\mathbf{u} = U^t\mathbf{x}$ states

$$u_1 = \frac{-x_1 + x_2}{\sqrt{2}} \text{ and } u_2 = \frac{x_1 + x_2}{\sqrt{2}}.$$

Substitute these values into Equation (8) in Example 4 and verify that the result is Equation (7).

6. Answer "True" or "False" and explain.

(a) If A is a 2×2 matrix with $det\ A < 0$, then $\mathbf{x}^t A\mathbf{x} = 1$ is a hyperbola.

(b) If A is a 2×2 matrix with $det\ A = 0$, then $\mathbf{x}^t A\mathbf{x} = 1$ is an ellipse.

(c) If A is a 3×3 matrix with $det\ A > 0$, then all eigenvalues of A must be positive.

7. How does the nature of the critical point of $f(x,y,z) = 2x^2 + 2\lambda xz + y^2 + 3z^2$ depend on λ?

8. Find and classify the critical points of $3x^2 - 4xy + 3y^2 + 8z^2$.

9. The point (0,0,0) is a critical point of $f(x,y,z) = x^2 + 2y^3 - 24yz + 16z^3$. Find the other critical point and classify both.

5.4. Systems of Differential Equations

We are going to use eigenvector-eigenvalue analysis to solve some systems of differential equations. One of the most important differential equations you have already studied is

$$y'(t) = ky(t). \tag{1}$$

Here, k is a constant, t is often thought of as time and $y(t)$ typically represents the amount of a substance present at time t or the size of a certain population at time t. The differential equation simply says that the rate of change of y at time t is proportional to the size of y at time t and the proportionality factor is independent of time. Perhaps, the most familiar examples of a process giving rise to this differential equation are continuously compounded interest where y is the amount of money earning interest and k is the interest rate; radioactive decay in which case k is negative; and unrestricted population growth. In all real examples we know the value of y at one point in time which we can take to be $t = 0$ by resetting our watches. That is, in addition to (1) we also have an "initial condition" given which we can assume is of the form

$$y(0) = y_0 \tag{2}$$

for some number y_0.

It is shown in Calculus that the solution to (1) is $y(t) = Ce^{kt}$ where C is an arbitrary constant. You can verify now that functions of this form do solve (1). With the initial condition (2) the arbitrary constant gets pinned down – it must be $y(0)$ or y_0. So the solution to the initial value problem (1)-(2) is

$$y(t) = y(0)\ e^{kt}. \tag{3}$$

When there are two quantities of interest, say two populations, there are two differential equations. For example, $x_1(t)$ might represent the population of a particular species at time t and $x_2(t)$ might represent the population of a different species at the same time. If the species have no interactions (like polar bears and monarch butterflies), then the populations would satisfy a very simple system of differential equations:

$$\begin{aligned} x_1'(t) &= k_1 x_1(t) \\ x_2'(t) &= k_2 x_2(t) \end{aligned} \tag{4}$$

perhaps with initial conditions $x_1(0)$ and $x_2(0)$. To solve the system (4) we just solve each equation separately.

If the populations under study interact with each other, then the equations are slightly more complicated. In the simplest case we would have a system of the form

$$\begin{aligned} x_1'(t) &= ax_1(t) + bx_2(t) \\ x_2'(t) &= cx_1(t) + dx_2(t) \end{aligned} \tag{5}$$

with initial conditions $x(0) = c_1$ and $y(0) = c_2$ where a, b, c, and d are real numbers. We can write (5) in matrix form as

$$\mathbf{x}'(t) = A\mathbf{x}(t) \tag{6}$$

or simply

$$\mathbf{x}' = A\mathbf{x}. \tag{6$'$}$$

To solve (6) we try to write the functions $x_1(t)$ and $x_2(t)$ as weighted sums of two independent functions, $u_1(t)$ and $u_2(t)$, which solve a diagonal system. Say

$$\begin{aligned} u_1'(t) &= \lambda u(t) \\ u_2'(t) &= \mu u_2(t) \end{aligned} \tag{7}$$

where

$$\begin{aligned} x_1(t) &= \alpha u_1(t) + \beta u_2(t) \\ x_2(t) &= \gamma u_1(t) + \delta u_2(t) \end{aligned} \tag{8}$$

where λ, μ, α, β, γ, and δ are real numbers to be determined by manipulating the equations (5), (7) and (8).

Now,

$$x_1'(t) = ax_1(t) + bx_2(t) = a\{\alpha u_1(t) + \beta u_2(t)\} + b\{\gamma u_1(t) + \delta u_2(t)\}$$
$$= u_1(t)\{a\alpha + b\gamma\} + u_2(t)\{a\beta + b\delta\}.$$

Also,

$$x_1'(t) = \alpha u_1'(t) + \beta u_2'(t) = \alpha\lambda u_1(t) + \beta\mu u_2(t).$$

Hence,

$$u_1(t)\{a\alpha + b\gamma\} + u_2(t)\{a\beta + b\delta\} = \alpha\lambda u_1(t) + \beta\mu u_2(t). \tag{9}$$

Since $u_1(t)$ and $u_2(t)$ are to be independent, it must follow that the multipliers of u_1 on each side of (9) must be equal. The same holds for the multipliers of $u_2(t)$. So we have the equations

$$a\alpha + b\gamma = \alpha\lambda \quad \text{and} \quad a\beta + b\delta = \beta\mu.$$

Repeating these manipulations using $x_2'(t)$ gives

$$x_2'(t) = cx_1(t) + dx_2(t) = c\{\alpha u_1(t) + \beta u_2(t)\} + d\{\gamma u_1(t) + \delta u_2(t)\}$$
$$= u_1(t)\{c\alpha + d\gamma\} + u_2(t)\{c\beta + d\delta\}$$

and

$$x_2'(t) = \gamma u_1'(t) + \delta u_2'(t) = \gamma\lambda u_1(t) + \delta\mu u_2(t).$$

And finally,

$$u_1(t)\{c\alpha + d\gamma\} + u_2(t)\{c\beta + d\delta\} = \gamma\lambda u_1(t) + \delta\mu u_2(t).$$

This gives two new relations

$$c\alpha + d\gamma = \gamma\lambda \quad \text{and} \quad c\beta + d\delta = \delta\mu.$$

The net result of the above manipulations is that if (5), (7), and (8) are to hold, then the following relations must be fulfilled:

$$\begin{aligned} a\alpha + b\gamma &= \alpha\lambda & \qquad a\beta + b\delta &= \beta\mu, \\ c\alpha + d\gamma &= \gamma\lambda & \qquad c\beta + d\delta &= \delta\mu. \end{aligned}$$

If we can write these in matrix-vector notation, their meaning becomes clear.

$$\begin{bmatrix} a & b \\ c & d \end{bmatrix}\begin{bmatrix} \alpha \\ \gamma \end{bmatrix} = \lambda\begin{bmatrix} \alpha \\ \gamma \end{bmatrix} \qquad \begin{bmatrix} a & b \\ c & d \end{bmatrix}\begin{bmatrix} \beta \\ \delta \end{bmatrix} = \mu\begin{bmatrix} \beta \\ \delta \end{bmatrix}. \tag{10}$$

The numbers λ and μ must be eigenvalues of A and in the relations

$$\begin{aligned} x_1(t) &= \alpha u_1(t) + \beta u_2(t) \\ x_2(t) &= \gamma u_1(t) + \delta u_2(t) \end{aligned}$$

$\begin{bmatrix} \alpha \\ \gamma \end{bmatrix}$ and $\begin{bmatrix} \beta \\ \delta \end{bmatrix}$ must be eigenvectors of A. If we call these eigenvectors $\mathbf{v}_1$ and $\mathbf{v}_2$ and let $\mathbf{x}(t)$ denote the vector $\begin{bmatrix} x_1(t) \\ x_2(t) \end{bmatrix}$, then the relation (8) reads

$$\mathbf{x}(t) = u_1(t)\mathbf{v}_1 + u_2(t)\mathbf{v}_2$$

where $u_1(t)$ and $u_2(t)$ are the solutions to (7). But we know how to solve (7): $u_1(t) = c_1 e^{\lambda t}$ and $u_2(t) = c_2 e^{\mu t}$ where c_1 and c_2 are arbitrary constants. So

$$\mathbf{x}(t) = c_1 e^{\lambda t}\mathbf{v}_1 + c_2 e^{\mu t}\mathbf{v}_2.$$

If the eigenvalues are complex, then the solution can be expressed in terms of sines and cosines. We will not treat this case in any detail. If A does not have two independent eigenvectors, then the method presented here does not give all solutions and must be modified.

***Theorem* 1.** If the 2×2 matrix A has two independent eigenvectors {this will be the case if A has two distinct eigenvalues}, then the *general solution* of the system

$$\mathbf{x}' = A\mathbf{x}$$

is

$$\mathbf{x}(t) = c_1 e^{\lambda t}\mathbf{v}_1 + c_2 e^{\mu t}\mathbf{v}_2$$

where λ and μ are the eigenvalues of A and $\mathbf{v}_1$ and $\mathbf{v}_2$ are the corresponding eigenvectors.

***Example* 1.** Solve the system $\mathbf{x}' = A\mathbf{x}$ where $x_1(0) = 1$ and $x_2(0) = 8$ and

$$A = \begin{bmatrix} 0 & 1 \\ 6 & 1 \end{bmatrix}.$$

For what initial conditions $x_1(0)$, $x_2(0)$ do the components of the solution remain bounded as t goes to infinity?

Solution. $$det(A - \lambda I) = det\begin{bmatrix} -\lambda & 1 \\ 6 & 1-\lambda \end{bmatrix} = \lambda(\lambda-1) - 6.$$

So the eigenvalues are 3 and −2. Solving $A\mathbf{v}_1 = 3\mathbf{v}_1$ with $\mathbf{v}_1 = \begin{bmatrix} r \\ s \end{bmatrix}$ gives

$$\begin{aligned} s &= 3r \\ 6r + s &= 3s. \end{aligned}$$

So the eigenvector has the form $\begin{bmatrix} r \\ 3r \end{bmatrix}$ for $r \neq 0$. A natural choice is $\mathbf{v}_1 = \begin{bmatrix} 1 \\ 3 \end{bmatrix}$. Similarly, an eigenvector for −2 is $\mathbf{v}_2 = \begin{bmatrix} 1 \\ -2 \end{bmatrix}$. So, in vector form the general solution is

$$\mathbf{x}(t) = c_1 e^{3t}\,\mathbf{v}_1 + c_2 e^{-2t}\mathbf{v}_2, \tag{11}$$

which is short hand for

$$x_1(t) = c_1 e^{3t} + c_2 e^{-2t}$$
$$x_2(t) = c_1 3 e^{3t} - c_2 2 e^{-2t}. \tag{12}$$

The initial conditions are used to determine c_1 and c_2:

$$1 = x_1(0) = c_1 e^0 + c_2 e^0 = c_1 + c_2$$
$$8 = x_2(0) = 3c_1 e^0 - 2c_2 e^0 = 3c_1 - 2c_2.$$

The unique solution is $c_1 = 2$, $c_2 = -1$. Hence, the solution to the system is

$$x_1(t) = 2e^{3t} - e^{-2t}$$
$$x_2(t) = 6e^{3t} - 2e^{-2t}.$$

Now, from (11) it follows that the components of the solution, $x_1(t)$ and $x_2(t)$, will become unbounded as t goes to infinity unless $c_1 = 0$. So, to have bounded solutions, the initial values must be such that c_1 is forced to be 0. Now notice that

$$\mathbf{x}(0) = c_1\mathbf{v}_1 + c_2\mathbf{v}_2.$$

Making $c_1 = 0$ is equivalent to having initial conditions $\mathbf{x}(0)$ proportional to $\mathbf{v}_2$. Therefore, the solutions $x_1(t)$ and $x_2(t)$ remain bounded as $t \to \infty$ if and only if $\mathbf{x}(0)$ is a multiple of $\mathbf{v}_2$, i.e., $x_2 = -2x_1(0)$. □

Example 2. (a) Find the general solution of the system $\{x$ and y are functions of $t\}$

$$x' = x + 2y$$
$$y' = 2x + y.$$

(b) Find the solution that satisfies $x(0) = 10$ and $y(0) = 6$.

(c) For what initial conditions $x(0)$ and $y(0)$ are both x and y bounded as t goes to infinity?

Solution. The matrix $\begin{bmatrix} 1 & 2 \\ 2 & 1 \end{bmatrix}$ has eigenvalues 3 and -1 with eigenvectors $\mathbf{v}_1 = \begin{bmatrix} 1 \\ 1 \end{bmatrix}$ and $\mathbf{v}_2 = \begin{bmatrix} -1 \\ 1 \end{bmatrix}$. So the general solution is $\mathbf{x}(t) = c_1 e^{3t}\mathbf{v}_1 + c_2 e^{-t}\mathbf{v}_2$. In terms of coordinate functions

$$x(t) = c_1 e^{3t} - c_2 e^{-t}$$

and

$$y(t) = c_1 e^{3t} + c_2 e^{-t}.$$

This solves part (a).

For part (b) we let $t = 0$ and obtain equations for c_1 and c_2

$$10 = x(0) = c_1 - c_2$$
$$6 = y(0) = c_1 + c_2.$$

The solution is $c_1 = 8$ and $c_2 = -2$ which gives the answer to part (b)

$$x(t) = 8e^{3t} + 2e^{-t}$$

and

$$y(t) = 8e^{3t} - 2e^{-t}.$$

For part (c) we observe that since e^{3t} grows unbounded and e^{-t} goes to zero as t goes to infinity, the only way we can have a bounded solution is if the coefficient, c_1, of e^{3t} is zero. Now c_1 and c_2, the coefficient of e^{-t}, satisfy the system

$$x(0) = c_1 - c_2$$
$$y(0) = c_1 + c_2.$$

If we are to have $c_1 = 0$, then we must have $x(0) = -c_2$ and $y(0) = c_2$ for an arbitrary c_2. In other words the *initial values of x and y must be negatives of each other. Any other choice of initial values gives rise to a solution that becomes unbounded as t goes to infinity.* Now notice that the initial vector

$\begin{bmatrix} x(0) \\ y(0) \end{bmatrix}$ gives a bounded solution if and only if it is a multiple of the eigenvector $\mathbf{v}_2$ (which is the eigenvector for -1). This is not an accident and could have been deduced as in the previous example by looking at the vector form of the general solution

$$\mathbf{x}(t) = c_1 e^{3t}\,\mathbf{v}_1 + c_2 e^{-t}\mathbf{v}_2.$$

As t varies, the components of $\mathbf{x}(t)$ have one component in the $\mathbf{v}_1$ direction and one in the $\mathbf{v}_2$ direction. At time t the component in the $\mathbf{v}_1$ direction is just $c_1 e^{3t}$ and the $\mathbf{v}_2$ component is $c_2 e^{-t}$. As t grows the $\mathbf{v}_1$ component will dominate unless c_1 is zero. If $c_1 = 0$, then $\mathbf{x}(t)$ is a multiple of $\mathbf{v}_2$ for all values of t. In particular when $t = 0$

$$\mathbf{x}(0) = c_2 e^{-0}\mathbf{v}_2 = c_2\mathbf{v}_2.$$

□

5.4. EXERCISES

1. The matrix $\begin{bmatrix} 5 & -1 \\ -1 & 5 \end{bmatrix}$ has eigenvectors $\begin{bmatrix} 1 \\ 1 \end{bmatrix}$ and $\begin{bmatrix} -1 \\ 1 \end{bmatrix}$. Solve the system

$$\begin{aligned} 5x(t) - y(t) &= x'(t) \\ -x(t) + 5y(t) &= y'(t) \end{aligned}$$

where $x(0) = 0$ and $y(0) = 20$.

2. Find the solution of the system of differential equations $\mathbf{x}'(t) = A\mathbf{x}(t)$ that satisfies the initial condition $x_1(0) = 100$, $x_2(0) = 50$ for the following choices of A.

(a) $\begin{bmatrix} -1 & -2 \\ -5 & 1 \end{bmatrix}$ (b) $\begin{bmatrix} 3 & -1 \\ -2 & 4 \end{bmatrix}$ (c) $\begin{bmatrix} -2 & 1 \\ 5 & 2 \end{bmatrix}$ (d) $\begin{bmatrix} 3 & -1 \\ -2 & 2 \end{bmatrix}$

3. Let the functions $x(t)$ and $y(t)$ solve the system

$$\begin{aligned} x'(t) &= -x(t) + y(t) \\ y'(t) &= \beta x(t) - 2y(t) \end{aligned}$$

with $x(0) = 10$ and $y(0) = 10$ where β is a parameter {so the solutions also depend on the value of β}. Find five different values of β for which the solutions satisfy

$$\lim_{t \to \infty} x(t) = \lim_{t \to \infty} y(t) = 0.$$

4. Solve the system of differential equations

$$x'(t) = -2x(t) + 4y(t)$$
$$y'(t) = x(t) - 2y(t)$$

with initial conditions $x(0) = 100$ and $y(0) = 300$.

5. The matrix $A = \begin{bmatrix} 3 & 1 \\ -1 & 1 \end{bmatrix}$ has only one independent eigenvector, $\mathbf{v} = \begin{bmatrix} 1 \\ -1 \end{bmatrix}$ with eigenvalue $\lambda = 2$. The steps below outline a method for solving $\mathbf{x}' = A\mathbf{x}$.

(a) Verify that for any vector $\mathbf{w}$ we have

$$A\mathbf{w} = \lambda\mathbf{w} + \beta\mathbf{v}$$

where β depends on only $\mathbf{w}$. What is β if $\mathbf{w} = \begin{bmatrix} 1 \\ 1 \end{bmatrix}$?

(b) Let $T = \begin{bmatrix} \lambda & \beta \\ 0 & \lambda \end{bmatrix}$ and $U = [\mathbf{v} \;\; \mathbf{w}]$. Show that $AU = UT$. So that $A = UTU^{-1}$.

(c) Let $\mathbf{z} = U^{-1}\mathbf{x}$. Show that $\mathbf{x}$ solves $\mathbf{x}' = A\mathbf{x}$ if and only if $\mathbf{z}$ solves $\mathbf{z}' = T\mathbf{z}$.

(d) The solution of $\mathbf{z}' = T\mathbf{z}$ can be found with techniques for first order, linear differential equations. Solve this system using the λ and β of part (a).

(e) Let $\mathbf{x} = U\mathbf{z}$ and verify that $\mathbf{x}$ solves the system $\mathbf{x}' = A\mathbf{x}$.

APPENDIX A The Jacobian Matrix and Linearization

Definition. A function that takes points in $\mathbf{R}^n$ to points in $\mathbf{R}^k$ is called a *vector-valued function of n-variables.*

The specification of such a function requires k-functions each of which is a real-valued function of n-variables.

Example 1. The polar coordinate change of variables $x = r \cos \theta$, $y = r \sin \theta$ can be studied in terms of the function

$$T(r,\theta) = (r \cos \theta,\ r \sin \theta) = \begin{bmatrix} r \cos \theta \\ r \sin \theta \end{bmatrix}$$

which takes the r-θ coordinate plane to x-y plane. In general changes of coordinate systems can be viewed in terms of vector-valued functions. □

Example 2. Consider $s(\theta,\phi) = (\cos \theta \sin \phi,\ \sin \theta \sin \phi,\ \cos \phi)$. As θ varies between 0 and 2π and ϕ varies from 0 to π, the values of $s(\theta,\phi)$ cover the sphere $x^2 + y^2 + z^2 = 1$. (See Appendix C, Example 5 for the details). In general functions from $\mathbf{R}^2$ to $\mathbf{R}^3$ can be thought of as parameterizing surfaces in $\mathbf{R}^3$. □

Example 3. Consider the system of equations

$$\begin{aligned} x^3 - 3xy^2 - x &= 1 \\ 3x^2y - y^3 - y &= -1 \end{aligned}$$

and define $f(x,y) = (x^3 - 3xy^2 - x - 1,\ 3x^2y - y^3 - y + 1)$. Clearly, solving the system is equivalent to finding all (x,y) for which $f(x,y) = (0,0)$. Any system of equations is equivalent to an equation involving a vector-valued function. This equivalence is used in Appendix B to derive a numerical method for finding approximate solutions to systems of equations.

Definition. Let $f\colon \mathbf{R}^n \to \mathbf{R}^k$ be $f(\mathbf{x}) = (f_1(\mathbf{x}), f_2(\mathbf{x}), \ldots, f_k(\mathbf{x}))$. The real-valued functions

$f_1(\mathbf{x}), f_2(\mathbf{x}), \ldots, f_k(\mathbf{x})$ are called the *coordinate functions of f.*

Definition. Let $f: \mathbb{R}^n \to \mathbb{R}^k$. If the coordinate functions of f are sufficiently differentiable, then the *Jacobian matrix of f* is the $k \times n$ matrix whose k^{th} row consists of the coordinate functions of the gradient of f_k. The Jacobian of f is denoted by J_f. If all functions in J_f are evaluated at the point $\mathbf{x}_0$, we write $J_f(\mathbf{x}_0)$.

Example 4. For $T(r,\theta) = (r\cos\theta,\ r\sin\theta)$, we have the coordinate functions $T_1(r,\theta) = r\cos\theta$ and $T_2(r,\theta) = r\sin\theta$.

Hence, the Jacobian matrix of T is

$$J_T = \begin{bmatrix} \dfrac{\partial T_1}{\partial r} & \dfrac{\partial T_1}{\partial \theta} \\ \dfrac{\partial T_2}{\partial r} & \dfrac{\partial T_2}{\partial \theta} \end{bmatrix} = \begin{bmatrix} \cos\theta & -r\sin\theta \\ \sin\theta & r\cos\theta \end{bmatrix}.$$

Also,

$$J_T(3,\pi) = \begin{bmatrix} -1 & 0 \\ 0 & -3 \end{bmatrix}.$$

Example 5. Let $f(x,y,z) = (x^2+y^3,\ xyz)$. Then,

$$J_f = \begin{bmatrix} 2x & 3y^2 & 0 \\ yz & xz & xy \end{bmatrix}.$$

Example 6. For $S(\theta,\phi) = (\cos\theta\sin\phi,\ \sin\theta\sin\phi,\ \cos\phi)$ we have

$$J_S = \begin{bmatrix} -\sin\theta\sin\phi & \cos\theta\cos\phi \\ \cos\theta\sin\phi & \sin\theta\cos\phi \\ 0 & -\sin\phi \end{bmatrix} \text{ and } J_S(\tfrac{\pi}{2},0) = \begin{bmatrix} 0 & 0 \\ 0 & 1 \\ 0 & 0 \end{bmatrix}$$

The Jacobian matrix of a vector-valued function is also called the derivative matrix. It is the natural extension of the derivative concept to vector-valued functions of several variables. The most important feature is stated below in Theorem 1.

Definition. Let $f\colon \mathbb{R}^n \to \mathbb{R}^k$ and let $\mathbf{x}_0$ be a given point in $\mathbb{R}^n$. The ***linearization of f at*** $\mathbf{x}_0$ is the function

$$L(\mathbf{x}) = f(\mathbf{x}_0) + J_f(\mathbf{x}_0)(\mathbf{x}-\mathbf{x}_0).$$

To be consistent with the matrix vector product, $J_f(\mathbf{x}_0)(\mathbf{x}-\mathbf{x}_0)$, we think of the values of f as column vectors. L is an affine function; it is sometimes called the ***best affine approximation*** to f at $\mathbf{x}_0$.

If f is a real-valued function of one variable, then its linearization at the point x_0 is

$$L(x) = f(x_0) + f'(x_0)(x-x_0).$$

In this case $L(x)$ is the function whose graph is the tangent line to the graph of f at $(x_0, f(x_0))$.

If f is a real-valued function of two variables, then the linearization of f at the point (x_0, y_0) is the function whose graph is the plane tangent to the graph of f at the point $(x_0, y_0, f(x_0, y_0))$.

Example 7. Let $f(x,y) = x^2 - 3y^3$. Then, $J_f = (2x, -9y^2)$ and at the point (5,4) we have the linearization

$$L(x,y) = f(5,4) + J_f(5,4)\begin{bmatrix} x-5 \\ y-4 \end{bmatrix}$$

or

$$L(x,y) = -167 + [10 \quad -144]\begin{bmatrix} x-5 \\ y-4 \end{bmatrix}$$

or

$$L(x,y) = -167 + 10(x-5) - 144(y-4).$$

Example 8. For $g(x,y) = (x^3 - 3xy^2 - x - 1,\ 3x^2y - y^3 - y + 1)$ we have

$$J_g = \begin{bmatrix} 3x^2 - 3y^2 - 1 & -6xy \\ 6xy & 3x^2 - 3y^2 - 1 \end{bmatrix},$$

$$J_g(1,2) = \begin{bmatrix} -10 & -12 \\ 12 & -10 \end{bmatrix}, \text{ and } g(1,2) = \begin{bmatrix} -13 \\ -3 \end{bmatrix}.$$

Hence, the linearization at (1,2) is

$$L(\mathbf{x}) = \begin{bmatrix} -13 \\ -3 \end{bmatrix} + \begin{bmatrix} -10 & -12 \\ 12 & -10 \end{bmatrix}\begin{bmatrix} x-1 \\ y-2 \end{bmatrix}, \text{ when } \mathbf{x} = \begin{bmatrix} x \\ y \end{bmatrix}.$$

Example 9. Let $f(x,y) = (x^2 - y^2 - 2,\ 2xy)$. The Jacobian matrix of f is

$$J_f = \begin{bmatrix} 2x & -2y \\ 2y & 2x \end{bmatrix}.$$

At the point (1.5, 1) we have

$$J_f(1.5,\ 1) = \begin{bmatrix} 3 & -2 \\ 2 & 3 \end{bmatrix} \text{ and } f(1.5,\ 1) = \begin{bmatrix} -0.75 \\ 3 \end{bmatrix}$$

and the linearization

$$L(x,y) = \begin{bmatrix} -0.75 \\ 3 \end{bmatrix} + \begin{bmatrix} 3 & -2 \\ 2 & 3 \end{bmatrix}\begin{bmatrix} x-1.5 \\ y-1 \end{bmatrix}.$$

Theorem 1. Let f: $\mathbb{R}^n \to \mathbb{R}^k$ and let x_0 be a point in $\mathbb{R}^n$. The linearization of f is the only affine function from $\mathbb{R}^n$ to $\mathbb{R}^k$ that satisfies

(a) $L(\mathbf{x}_0) = f(\mathbf{x}_0)$

and (b) $\lim\limits_{\mathbf{x} \to \mathbf{x}_0} \dfrac{|L(\mathbf{x}) - f(\mathbf{x})|}{|\mathbf{x}-\mathbf{x}_0|} = 0.$

Any affine mapping of the form $g(\mathbf{x}) = f(\mathbf{x}_0) + B(\mathbf{x}-\mathbf{x}_0)$ where B is an $k \times n$ matrix satisfies (a). Condition (b) requires that $L(\mathbf{x})$ be a very good approximation to $f(\mathbf{x})$ if $\mathbf{x}$ is close to $\mathbf{x}_0$ — the point of linearization. In fact for $\mathbf{x}$ "close enough" to $\mathbf{x}_0$ the distance between $L(\mathbf{x})$ and $f(\mathbf{x})$ must be much smaller than the distance between $\mathbf{x}$ and $\mathbf{x}_0$. The term "close enough" cannot be quantified without detailed knowledge of the second order partials of f. This lack of precision will not be an impediment for using $L(\mathbf{x})$ as an approximation to $f(\mathbf{x})$ for values of $\mathbf{x}$ near $\mathbf{x}_0$.

EXERCISES

1. Find the Jacobian matrix of the following functions and evaluate it at the given point.
   (a) $f(x,y) = (x+y-2,\ x^2+xy+y^2)$ at $\mathbf{x}_0 = (2,0)$.
   (b) $g(x,y) = (x^2+y^2-1,\ x^2-y^2)$ at $\mathbf{x}_0 = (0.4,\ 1.2)$.
   (c) $h(x,y) = (e^x+xy-1,\ \sin(xy)+x+y-1)$ at $\mathbf{x}_0 = (0.1,\ 0.5)$.
   (d) $f(x) = (\cos x,\ 2 \sin x,\ x)$ at $x = \pi$.
   (e) $f(x,y) = (xy,\ \cos x+\sin y,\ x^2y^2)$ at $(1,0)$.
   (f) $g(x,y,t) = (xyt,\ x^2+xyt+t^2)$ at $\mathbf{x}_0 = (2,\ 1,\ -3)$.

2. For each function in problem 1 find the linearization at the given point.

3. Let $f(x,y)$ be the function of Example 9. And let $L(x,y)$ be the linearization of f at the point $(1.5,1)$. Solve the linear system $L(x,y) = (0,0)$.

Appendix B Newton's Method for Non-Linear Systems

We now apply the concept of linearization to the problem of solving non-linear systems of the form

$$f(x) = 0 \tag{1}$$

where $f: \mathbb{R}^n \to \mathbb{R}^n$ and $\mathbf{0}$ is the zero vector in $\mathbb{R}^n$. As we mentioned in Appendix A every system can be written this way.

If we believe that a particular point x_0 is close to a solution of (1), we can replace the equation $f(\mathsf{x}) = \mathbf{0}$ by $L(\mathsf{x}) = \mathbf{0}$ where L is the linearization of f at x_0. The system that results is then linear and we can use Gaussian Elimination (or some other correct method) to solve it. This new, *linearized system* has the form

$$J_f(\mathsf{x}_0)(\mathsf{x}-\mathsf{x}_0) = -f(\mathsf{x}_0). \tag{2}$$

Symbolically the solution, which we'll denote by x_1, is

$$\mathsf{x}_1 = \mathsf{x}_0 - [J_f(\mathsf{x}_0)]^{-1} f(\mathsf{x}_0).$$

The formula for x_1 is quite similar to the one-dimensional Newton iteration formula

$$x_1 = x_0 - f'(x_0)^{-1} f(x_0). \tag{3}$$

And, as with the one-dimensional Newton iteration, we may repeat the process using the newly found x_1 in the place of x_0. Here is the general algorithm.

Newton's Method for Non-Linear Systems

Given $f: \mathbb{R}^n \to \mathbb{R}^n$, to approximately solve $f(\mathsf{x}) = \mathbf{0}$:

1. Make an initial guess, x_{old}, at the solution and set the iteration counter $N = 0$.
2. Compute $J_f(\mathsf{x}_{old})$ and $f(\mathsf{x}_{old})$ and set $N = N+1$.
3. Solve $J_f(\mathsf{x}_{old})\mathsf{y} = f(\mathsf{x}_{old})$.
4. Let $\mathsf{x}_{new} = \mathsf{x}_{old} - \mathsf{y}$.
5. If x_{new} is an acceptable solution, stop; otherwise define $\mathsf{x}_{old} = \mathsf{x}_{new}$ and restart at step 2.

Remarks. 1. If f is sufficiently well behaved and if the initial guess is sufficiently close to a solution of $f(\mathsf{x}) = \mathbf{0}$, then Newton's method will converge very rapidly to this solution.

2. There is no universally good strategy for choosing the initial guess. In general one must make use of some knowledge of $f(\mathbf{x})$. Exercises 5 and 6 indicate how sensitive the convergence can be to the choice of initial guess.

3. If the functions in J_f are very complicated, then sometimes one will modify step 2 and not compute a new $J_f(\mathbf{x}_{\text{old}})$ at every step. See Exercises 7 and 8, for example.

4. There are several criteria that are used to determine if $\mathbf{x}_{\text{new}}$ is an acceptable solution. For example, if the magnitude of the vector $f(\mathbf{x}_{\text{new}})$ is smaller than some predetermined tolerance, then one might stop on the grounds that $\mathbf{x}_{\text{new}}$ nearly solves the system. Or if $|\mathbf{x}_{\text{new}}-\mathbf{x}_{\text{old}}|$ is sufficiently small, one could stop on the grounds that little improvement is being made. In any case, one should stop if the iteration counter N becomes too big.

Example 1. Use Newton's Method to find a solution of

$$\begin{aligned} x_1^2 + x_2^2 &= 1 \\ x_1^2 - x_2^2 &= -0.5 \end{aligned}$$

in the first quadrant.

Solution. We have

$$f(x_1,x_2) = \begin{bmatrix} x_1^2 + x_2^2 - 1 \\ x_1^2 - x_2^2 + 0.5 \end{bmatrix} \text{ and } \mathrm{J}_f = \begin{bmatrix} 2x_1 & 2x_2 \\ 2x_1 & -2x_2 \end{bmatrix}.$$

We'll take $\mathbf{x}_{\text{old}} = \begin{bmatrix} 1 \\ 3 \end{bmatrix}$ so $f\begin{bmatrix} 1 \\ 3 \end{bmatrix} = \begin{bmatrix} 9 \\ -7.5 \end{bmatrix}$ and $J_f(1,3) = \begin{bmatrix} 2 & 6 \\ 2 & -6 \end{bmatrix}$. The system $J_f(1,3)\mathbf{y} = f(1,3)$ is

$$\begin{aligned} 2y_1 + 6y_2 &= 9 \\ 2y_1 - 6y_2 &= -7.5 \end{aligned}$$

and has solution $y_1 = 0.375$, $y_2 = 1.375$. This gives

$$\mathbf{x}_{\text{new}} = \mathbf{x}_{\text{old}} - \mathbf{y} = (0.625,\ 1.625)$$

Continuing this process using $\mathbf{x}_{\text{new}}$ to be our new $\mathbf{x}_{\text{old}}$, we generate the following table.

Table B-1

N	x_1	x_2
0	1	3
1	0.735	1.375
2	0.512	1.04
3	0.5001	0.88108
4	0.50000002	0.86615404
6	0.5	0.86602540

The exact solution is $x_1 = 0.5$ and $x_2 = \frac{\sqrt{3}}{2} = 0.866025404...$ Newton's method gave a good approximation after only 4 steps and an excellent approximation after 6 steps. □

Example 2. We'll try to find a solution of the system

$$x_1^2 + (1.01)x_2^2 = 0.98$$

$$x_1^2 \sin(0.1)x_2 - x_2^2 = -0.5$$

in the first quadrant.

Solution. This system bears a strong resemblance to the system in Example 1. The coefficients in the first equations are very close to each other. The second equations differ in only the $\sin(0.1)x_2$ term. Since $\sin(0.1)x_2$ is fairly small if x_2 is close to $\frac{\sqrt{3}}{2}$, the x_2 coordinate of the solution of the first system, we feel that this term will have only a little influence on the solution. Therefore, as our initial guess we take $\mathbf{x}_{\text{old}}$ to be the solution of Example 1. Now, we have

$$f(x_1, x_2) = \begin{bmatrix} x_1^2 + (1.01)x_2^2 - 0.98 \\ x_1^2 \sin(0.1)x_2 - x_2^2 + 0.5 \end{bmatrix}$$

$$\text{and } J_f = \begin{bmatrix} 2x_1 & 2.02x_2 \\ 2x_1 \sin(0.1)x_2 & 0.1x_1^2 \cos(0.1)x_2 - 2x_2 \end{bmatrix}$$

The following results were obtained with the aid of a computer program.

Table B-2.

N	x_1	x_2	$f(x_1,x_2)$
0	0.5	0.86602540	(0.275, −0.2283764)
1	0.6897106	0.7418604	(0.05156, −0.01509)
2	0.6656155	0.7296327	(0.0007316, −0.0000667)
3	0.6651532	0.7295539	(0.0000002, −0.0000000)
4	0.6651530	0.7295539	$(-1.3\times10^{-8}, -2.5\times10^{-8})$

From the third to the fourth iterate only the seventh decimal place of x_1 changes. Our initial guess was excellent! □

Here is the main theorem on the convergence of Newton's Method. The key to a successful implementation is the initial guess.

Theorem. Let f: $\mathbb{R}^n \to \mathbb{R}^n$ be sufficiently differentiable and let $\mathbf{x}^*$ be a solution of $f(\mathbf{x}) = 0$. If $J_f(\mathbf{x}^*)$ is invertible and if the initial guess $\mathbf{x}_0$ is sufficiently close to $\mathbf{x}^*$, then Newton's Method converges to the solution $\mathbf{x}^*$. Furthermore, the error at step n, $|\mathbf{x}^* - \mathbf{x}_n|$ is proportional to the square of the error at step $n-1$ for sufficiently large n.

Roots of Polynomials

Complex roots of polynomial equations can be found with Newton's method. An equation of the form

$$f(z) = 0$$

where $z = x + iy$ and $i^2 = -1$ can be written as a system of two equations (one involving the real part and one involving the imaginary part) in the two unknowns x and y. We illustrate this with the equation

$$z^3 - z - c = 0.$$

First, $z = x + iy$ so

$$\begin{aligned} z^3 = (x+iy)^3 &= x^3 + 3x^2iy + 3xi^2y^2 + i^3y^3 \\ &= (x^3 - 3xy^2) + i(3x^2y - y^3). \end{aligned}$$

If $c = a + ib$, where a and b are real, then the original equation can be written as

$$(x^3 - 3xy^2 - x - a) + i(3x^2y - y^3 - y - b) = 0.$$

This is really two equations, one for the real part and one for the imaginary part. That is, the original complex variable equation is equivalent to the following system of two equations involving two real variables:

$$\begin{aligned} x^3 - 3xy^2 - x - a &= 0 \\ 3x^2y - y^3 - y - b &= 0. \end{aligned}$$

It is a straightforward task to apply Newton's Method to this system.

Example 3. Find a solution of $z^2 = i+2$ where $i^2 = -1$.
Solution. We have $f(z) = z^2 - i - 2$ in terms of complex numbers or

$$f(x+iy) = (x + iy)^2 - i - 2 = x^2 - y^2 - 2 + i(2xy - 1)$$

Therefore, in terms of real variables we have

$$\begin{aligned} x^2 - y^2 - 2 &= 0 \\ 2xy - 1 &= 0. \end{aligned}$$

So, with

$$f(x,y) = \begin{bmatrix} x^2 - y^2 - 2 \\ 2xy - 1 \end{bmatrix} \text{ and } J_f = \begin{bmatrix} 2x & -2y \\ 2y & 2x \end{bmatrix}$$

and initial guess $\mathbf{x}_{\text{old}} = (1,0)$ we get

Table B-3

N	x	y	$f(x,y)$
0	1	0	$(-1, -1)$
1	1.5	0.5	$(0, 0.5)$
2	1.45	0.35	$(-0.1999999, 0.015)$
3	1.455337	0.3435393	$(-0.0000133, -0.000069)$
4	1.4553467	0.3435608	$(3.5\times10^{-8}, 2.5\times10^{-8})$

With the initial guess (0,1) it takes seven iterations to achieve this accuracy. If we made the initial guess (0,0), however, we would fail at step 1 since $J_f(0,0)$ is the zero matrix. □

Computer studies of the dependence of the initial guess on the convergence of Newton's Method applied to complex polynomials have helped make the subject of Fractal Geometry well known. Exercises 5 and 6 are related to this study.

A Geometric Interpretation

For the system

$$f_1(x,y) = 0$$
$$f_2(x,y) = 0$$

with initial guess (x_0,y_0) Newton's Method gives the following linear system (cf. (2))

$$\frac{\partial f_1}{\partial x}(x_0,y_0)(x-x_0) + \frac{\partial f_1}{\partial y}(x_0,y_0)(y-y_0) + f_1(x_0,y_0) = 0$$

$$\frac{\partial f_2}{\partial x}(x_0,y_0)(x-x_0) + \frac{\partial f_2}{\partial y}(x_0,y_0)(y-y_0) + f_2(x_0,y_0) = 0.$$

The points (x,y) that satisfy the first of these are the points on the plane which is tangent to the graph of $f_1(x,y)$ at $(x_0,y_0,f_1(x_0,y_0))$ and which lie in the $x-y$ plane. Similarly, the solution set of the second equation is the intersection of the $x-y$ plane with a tangent plane to the graph of $f_2(x,y)$. So, a step of Newton's Method for two equations in two unknowns consists in finding the intersection of 3 planes: the $x-y$ plane and one tangent plane to each of the coordinate functions.

Higher Order Systems

The geometric description of Newton's Method for systems with more than two unknowns is not easy to visualize. However, the algebraic computations are easy to set up and their solution is — in principle — straightforward.

EXERCISES

1. Compute one iteration of Newton's Method for each of the following systems using the given initial guess.

(a) $x + y = 2$, $x^2 + xy + y^2 = 5$, $\mathsf{x}_0 = (2,0)$

(b) $x^2 + y^2 = 1$, $x^2 - y^2 = 0.5$, $\mathsf{x}_0 = (-0.45,\ 1.2)$

(c) $x^2 - y^2 - x = -1$, $2xy - y = 0$, $\mathsf{x}_0 = (0.5,\ 0.8)$

2. In Example 1 for what set of initial guesses will Newton's method fail at the very first step?

3. Use Newton's Method to find solutions to the system below. Use $\mathsf{x}_0 = (3.4,\ 2.2)$ and $\mathsf{z}_0 = (1,-2)$ as initial guesses and see what solutions you get.

$$x + 3\ \ell n\ x - y^2 = 0$$

$$2x^2 - xy - 5x + 1 = 0$$

4. Find the solutions of $z^3 - z = 1$.

5. The equation $z^3 = 1$ has three roots 1 and $(-\frac{1}{2}) \pm (\frac{\sqrt{3}}{2})$. Here is a way to study how the initial guess determines the convergence of Newton's Method. Select a portion of the plane say $a \le x \le b,\ c \le y \le d$; for each point in this set run Newton's Method for $z^3 - 1 = 0$ until it gets "close" to one of the roots or until you've computed more than a prescribed number of

iterations. If you converge to 1 with the initial guess (x,y), then color the initial guess your favorite color; if you converge to $(-\frac{1}{2}) + (\frac{\sqrt{3}}{2})i$, then color the initial guess your second favorite color; if you converge to $(-\frac{1}{2}) - (\frac{\sqrt{3}}{2})i$, then color your initial guess your third favorite color. If the iteration fails to converge after the prescribed number of iterations, don't color the point or color it a fourth color. Computer implementation of this idea requires, of course, that you break the portion of the plane $a \le x \le b$, $c \le y \le d$ into a finite number of boxes, each box corresponding to a point (pixel) on your computer display.

6. Repeat the experiment described above for the equations

   (a) $z^3 - z = 0$ {roots are 0, 1, −1}

   (b) $z^4 - 1 = 0$ {roots are 1, −1, i, −i -- you need five colors}

7. Newton's method for computing $\sqrt{2}$ uses the function $f(x) = x^2 - 2$. The iteration is

(a)
$$x_{new} = x_{old} - \frac{f(x_{old})}{f'(x_{old})}.$$

Consider the *modified* iteration step

(b)
$$x_{new} = x_{old} - \frac{f(x_{old})}{3}.$$

The denominator 3 is simply $f'(1.5)$. The modified iterated does not require the computation of f' at every step. Compare (a) and (b) for various choices of initial guesses.

8. For the system in Example 2 replace the system

(a)
$$J_f(x_{old})y = f(x_{old})$$

by

(b)
$$J_f(0.5,\ 0.866)y = f(x_{old}).$$

The advantage of (b) over (a) is that Gaussian Elimination has to be done only *once* on

$J_f(0.5,\ 0.866)$ and the information can be saved (see Section 1.3) whereas in (a) the elimination algorithm must be implemented for a new matrix at each iteration. Do we still converge? Rapidly? What happens if we use

$$J_f(1,3)\mathbf{y} = f(\mathbf{x}_{\text{old}})$$

at each step?

9. Let $f(x,y) = (x^2 - y^2 + 1,\ 2xy)$. Show that if the initial guess has the form $\mathbf{x} = (a,0)$, then the Newton iteration fails to converge. As initial guess of the form (a,b) gives convergence to $(0,1)$ if $b > 0$ and to $(0,-1)$ if $b < 0$.

10. One can apply the one dimensional Newton iteration to functions of complex variables simply by replacing the x's in equation (3) by z's. Show that this is equivalent to the approach presented here.

APPENDIX C Parametric Surfaces and Their Tangent Planes

As we mentioned in Example 2 of Appendix A we can think of a function that maps $\mathbb{R}^2$ to $\mathbb{R}^3$ as parameterizing a surface $\mathbb{R}^3$. This is similar to the way that we view functions from $\mathbb{R}$ to $\mathbb{R}^3$ as parameterizing curves in $\mathbb{R}^3$. A general function f: $\mathbb{R}^2 \to \mathbb{R}^3$ has the form

$$f(u,v) = (f_1(u,v); f_2(u,v), f_3(u,v))$$

where the f_i's are real valued functions of u and v. So, a point (x,y,z) is on the surface parameterized by f if and only if there are values of u and v so that simultaneously $x = f_1(u,v)$, $y = f_2(u,v)$, and $z = f_3(u,v)$.

Example 1. $F(u,v) = (2u-v+1,\ u,\ u-2v+3)$ parameterizes a plane. The plane passes through (1,0,3) and has normal vector $(2,1,1)\times(-1,0,-2)$.

Example 2. $g(s,t) = (s,\ t,\ s^2-t^2)$ parameterizes the surface $z = x^2-y^2$.

Example 3. $h(u,v) = (u \cos v,\ u \sin v,\ u)$ gives the cone whose Cartesian form is $z^2 = x^2 + y^2$. To verify this first note that $u^2 = (u \cos v)^2 + (u \sin v)^2$ holds for all values of u and v. This means that every point parameterized by h lies on the cone given by $z^2 = x^2 + y^2$. On the other hand, if (x,y,z) lies on this cone, and if we take u to be z, we can determine a unique value of v in the range $[0,2\pi]$ so that $x = u \cos v$ and $y = u \sin v$.

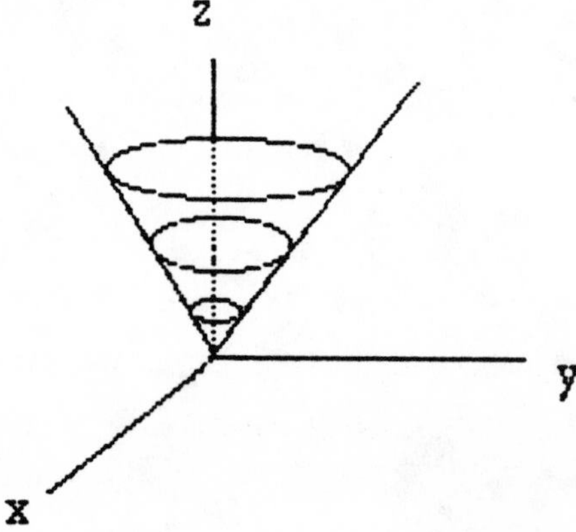

$$z^2 = x^2 + y^2$$
$$z \geq 0$$

The portion for $z < 0$ lies below the x-y plane. The intersection of the sketched surface with the y-z plane is the curve $z = |y|$.

Example 4. If $R > 1$, then as u and v vary between 0 and 2π the parametric equations

$$x = (R + \cos v)\cos u$$
$$y = (R + \cos v)\sin u$$
$$z = \sin v$$

give a torus {a doughnut}. If v is held fixed, say $v = a$, and u is allowed to vary, then the above equations describe a curve that is $\sin a$ units "above" the x-y plane. The x and y coordinates of points on this curve are given by

$$x = (R + \cos a)\cos u$$
$$y = (R + \cos a)\sin u.$$

As u varies between 0 and 2π, these equations describe a circle of radius $R + \cos a$. In summary, for fixed v the given parametric equations trace out a circle of radius $R + \cos a$; the center of the circle is $(0, 0, \sin a)$ and the circle is parallel to the x-y plane.

If u is held fixed, say $u = b$, and v is allowed to vary, then we have the parametric curve

$$x = (R + \cos v)\cos b$$
$$y = (R + \cos v)\sin b$$
$$z = \sin v.$$

Every point on this curve satisfies the equation $(\sin b)x - (\cos b)y = 0$ which is the equation of a plane passing through the origin. Therefore, this curve lies in the plane through the origin whose normal vector is $(\sin b, -\cos b, 0)$. Finally, we will show that points on this curve all lie one unit away from the point $(R \cos b, R \sin b, 0)$. This will show that our curve is the circle of radius 1 and center $(R \cos b, R \sin b, 0)$ in the plane $(\sin b)x - (\cos b)y = 0$. The required computation is

$$|(x-R \cos b,\ y-R \sin b,\ z)|^2 = (\cos v \cos b)^2 + (\cos v \sin b)^2 + (\sin v)^2 = 1.$$

Example 5. The sphere with center (0,0,0) and radius R is parameterized by

$$x = R \cos u \sin v$$
$$y = R \sin u \sin v$$
$$z = R \cos v$$

where $0 \le u \le 2\pi$ and $0 \le v \le \pi$. This is a direct consequence of the formulas for expressing points in space in terms of spherical coordinates. If (x,y,z) is a point on the sphere of radius R with center (0,0,0), then

$$x^2 + y^2 + z^2 = R^2.$$

From the fact that

$$(R \cos u \sin v)^2 + (R \sin u \sin v)^2 + (R \cos v)^2 = R^2$$

it follows that the given equations parameterize a *subset* of this sphere of radius R with center (0,0,0). The projection of a point (x,y,z) on this sphere onto the x-y plane is, of course, $(x,y,0)$. The polar coordinates of this point are $x = r \cos u$, $y = r \sin u$ where $r^2 = x^2 + y^2$. What is r? View the point (x,y,z) as being decomposed into a component parallel to the x-y plane and a component parallel to the z-axis. The first component has length r and the second has length $|z|$. If v is the angle that the line segment from (0,0,0) to (x,y,z) makes the z-axis, then from trigonometry we have $\sin v = \frac{r}{R}$ and $\cos v = \frac{z}{R}$. This gives $r = R \sin v$ and $z = R \cos v$. {Notice that when $z < 0$ the cosine is negative.}

$$R^2 = r^2 + z^2$$
$$r^2 = x^2 + y^2$$

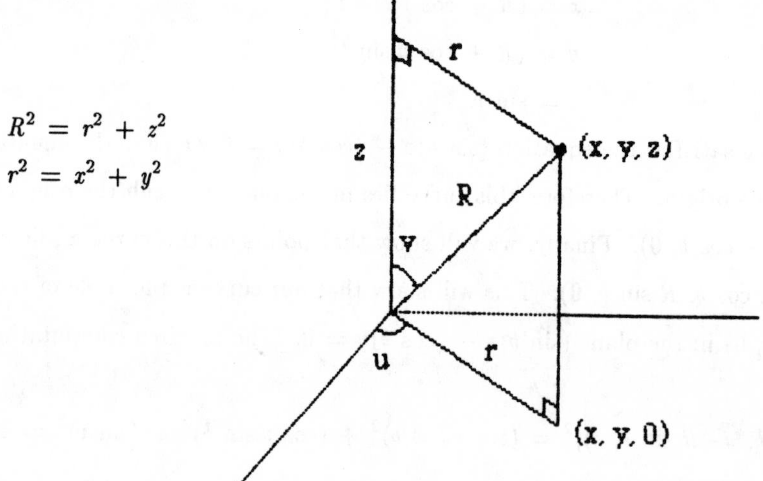

In general, if $f\colon \mathbb{R}^2 \to \mathbb{R}^3$ is given by

$$f(u,v) = (f_1(u,v), f_2(u,v), f_2(u,v)),$$

then the linearization of f at (u_0,v_0) has the form

$$L(u,v) = f(u_0,v_0) + J_f(u_0,v_0) \begin{bmatrix} u - u_0 \\ v - v_0 \end{bmatrix}$$

and the Jacobian of f at (u_0,v_0) has the form

$$J_f(u_0,v_0) = \begin{bmatrix} \frac{\partial f_1}{\partial u}(u_0,v_0) & \frac{\partial f_1}{\partial v}(u_0,v_0) \\ \frac{\partial f_2}{\partial u}(u_0,v_0) & \frac{\partial f_2}{\partial v}(u_0,v_0) \\ \frac{\partial f_3}{\partial u}(u_0,v_0) & \frac{\partial f_3}{\partial v}(u_0,v_0) \end{bmatrix}.$$

The rows of the Jacobian are the gradients of the coordinate functions. We'll now discuss the geometric meaning of the columns of the Jacobian.

If we force v to be a fixed value, say v_0, then as u varies $f(u,v_0)$ traces out a curve in space. The parametric equations of this curve are

$$x = f_1(u,v_0),\ y = f_2(u,v_0),\ z = f_3(u,v_0).$$

So, the velocity vector is

$$\begin{bmatrix} \frac{\partial f_1}{\partial u}(u,v_0) \\ \frac{\partial f_2}{\partial u}(u,v_0) \\ \frac{\partial f_3}{\partial u}(u,v_0) \end{bmatrix}.$$

At the value $u = u_0$ this is exactly the first column of the Jacobian of f at (u_0,v_0). Similarly, if we fix $u = u_0$ and let v vary, we get another curve whose velocity vector at $v = v_0$ is just the second column of the Jacobian. Each of the curves generated by fixing one of the parameters and varying the other must lie on the surface parameterized by f and both must pass through the point $f(u_0,v_0)$. Since the velocity vectors at this point are both tangent to surface, they determine a unique plane at this point – the tangent plane to the surface – provided they are not multiples of each other. The cross product of the two velocity vectors is a normal vector for the tangent plane.

Summary

Given the parametric surface $(x,y,z) = f(u,v)$ and the point (u_0,v_0) in "parameter space," the tangent plane to the surface at the point $f(u_0,v_0)$ can be found as follows:

1. Find $J_f(u_0,v_0)$ and call the first column $\mathbf{c}_1$ and the second $\mathbf{c}_2$.

2. The normal to the tangent plane to the surface at this point is $\mathbf{n} = \mathbf{c}_1 \times \mathbf{c}_2$ and $\mathbf{x} = (x,y,z)$ is on this tangent plane if and only if $\mathbf{n}^t(\mathbf{x}-f(u_0,v_0)) = 0$.

In short, the tangent plane at $f(u_0,v_0)$ is parameterized by the linearization of f at (u_0,v_0).

Example 6. We will find the equation of the tangent plane to the surface $x = u^2$, $y = uv$, $z = u+v$ at the point (1,2,3).

First we find the values of u and v that give the point (1,2,3). We have $1 = u^2$, $2 = uv$, and $3 = u+v$. The only solution is $u = 1$ and $v = 2$. This means that $f(1,2) = (1,2,3)$. So,

$$J_f = \begin{bmatrix} 2u & 0 \\ v & u \\ 1 & 1 \end{bmatrix} \quad \text{and} \quad J_f(1,2) = \begin{bmatrix} 2 & 0 \\ 2 & 1 \\ 1 & 1 \end{bmatrix}$$

The cross product of the columns of the last matrix is a normal to the desired plane. Since $(2,2,1)\times(0,1,1) = (1,-2,2)$, the tangent plane equation is $(1,-2,2)\cdot(x-1,y-2,z-3) = 0$ or $x-1-2(y-2) + 2(z-3) = 0$.

EXERCISES

1. Let $f(u,v) = ((2 + \cos v)\cos u, (2 + \cos v)\sin u, \sin v)$ for $0 \le u,v \le 2\pi$. Sketch the following curves.

(a) $f(u,0)$, $0 \le u \le 2\pi$

(b) $f(u,\pi)$, $0 \le u \le 2\pi$

(c) $f(0,v)$, $0 \le v \le 2\pi$

(d) $f(\frac{\pi}{2},v)$, $0 \le v \le 2\pi$

(Compare to Example 4 above.)

2. Show that the parametric curve $x = 4 \cos 2t$, $y = 3 \cos 2t$, $z = 5 \sin 2t$ lies on a sphere. What is the radius of the sphere?

3. Find the equation of the tangent plane to the given surface at the point determined by the given parameter values

(a) $x = u^2$, $y = v^2$, $z = u^2 + v^2$ for $u = 1$, $v = 1$

(b) $z = 3x^2 + 8xy$ for $x = 1$, $y = 0$

(c) $x = u^3$, $y = v^2$, $z = u + v$ for $u = 1$, $v = -1$.

4. Find the equation of the tangent plane to the given parametric surface at the given point

(a) $x = 2u$, $y = u^2 + v$, $z = v^2$ at $(0,1,1)$

(b) $x = u^2 - v^2$, $y = u + v$, $z = u^2 + 4v$ at $(-\frac{1}{4}, \frac{1}{2}, 2)$.

5. Suppose $f\colon \mathbb{R}^2 \to \mathbb{R}^3$ parameterizes a surface and that $f(2,3) = (7,8,9)$ and

$$J_f(2,3) = \begin{bmatrix} 1 & 3 \\ 0 & 3 \\ 2 & 3 \end{bmatrix}$$

Find the equation of the plane tangent to the surface at $(7,8,9)$.

6. Verify the statement that the linearization of f at (u_0,v_0) parameterizes the tangent plane at $f(u_0,v_0)$ to the surface traced out by f.

APPENDIX D Making Fractals with Affine Functions

In this recreational appendix we describe a technique based on iteration of collections of affine functions for generating exotic subsets of $\mathbb{R}^2$ called *fractals.* The geometric properties of the affine functions determine the shape of the generated set. Extensions of the method presented here show great promise as data reduction schemes for digitized images.

First, we consider the simple affine function

$$w(\mathsf{x}) = \alpha(\mathsf{x}-\mathsf{p}) + \mathsf{p} \qquad \square$$

where $0 < \alpha < 1$ and p is in $\mathbb{R}^2$. Notice that

$$w(\mathsf{p}) = \mathsf{p};$$

for this reason, p is called a *fixed point* of w. We can rewrite $w(\mathsf{x})$ as $w(\mathsf{x}) = \alpha\mathsf{x} + (1-\alpha)\mathsf{p}$ which shows that $w(\mathsf{x})$ lies on the line segment between x and p. So, $w(w(\mathsf{x}))$ is between $w(\mathsf{x})$ and p. Since $w(\mathsf{p}) = \mathsf{p}$ and affine functions preserve proportions, we have

$$\alpha = \frac{|w(\mathsf{x})-\mathsf{p}|}{|\mathsf{x}-\mathsf{p}|} = \frac{|w(w(\mathsf{x}))-w(\mathsf{p})|}{|w(\mathsf{x})-\mathsf{p}|} = \frac{|w(w(\mathsf{x}))-\mathsf{p}|}{\alpha|\mathsf{x}-\mathsf{p}|}.$$

So,

$$|w(w(\mathsf{x}))-\mathsf{p}| = \alpha^2|\mathsf{x}-\mathsf{p}|.$$

If we continue this process and define, $\mathsf{x}_1 = w(\mathsf{x})$, $\mathsf{x}_2 = w(\mathsf{x}_1),\ldots,\mathsf{x}_n = w(\mathsf{x}_{n-1}),\ldots$, we will get

$$|\mathsf{x}_n-\mathsf{p}| = \alpha^n|\mathsf{x}-\mathsf{p}|.$$

This shows that $\mathsf{x}_n \to \mathsf{p}$ as $n \to \infty$. More generally, one has

Theorem 1. If $w(\mathsf{x}) = A(\mathsf{x}-\mathsf{p}) + \mathsf{p}$ where p is in $\mathbb{R}^2$ and A is a 2×2 matrix whose eigenvalues have absolute value strictly less than one, then the sequence defined by x_0 = arbitrary, $\mathsf{x}_1 = w(\mathsf{x}_0),\ldots,\mathsf{x}_n = w(\mathsf{x}_{n-1})$ converges to p.

There are several natural ways to extend the notion of iteration to collections of affine functions. We will present two methods which for our purposes are equivalent. We start with a definition.

Definition. An *iterated function system* (IFS) is a finite collection of affine mappings $\{w_1(\mathsf{x}),...,w_d(\mathsf{x})\}$ from $\mathbb{R}^2$ to itself.

Set-Valued Iteration

We associate to the IFS $\{w_1,...,w_d\}$ the function W that turns sets into sets by the rule

$$W(S) = w_1(S) \cup w_2(S) \cup \cdots \cup w_d(S),$$

where $w_k(S) = \{w_k(\mathsf{x}): \mathsf{x} \in S\}$ is the image of S under w_k. We then define the iteration

$$S_n = W(S_{n-1})$$

where S_0 is an arbitrary closed, bounded subset of $\mathbb{R}^2$.

Example 1. Consider the IFS given by the function $w_1(\mathsf{x}) = \frac{1}{2}\mathsf{x}$, $w_2(\mathsf{x}) = \frac{1}{2}(\mathsf{x} - (1,0)) + (1,0)$. Starting with $S_0 = \{\mathsf{x}: |x_1| \leq 2 \text{ and } |x_2| \leq 2\}$, and noticing that $w_2(\mathsf{x}) = \frac{1}{2}\mathsf{x} + (\frac{1}{2},0)$, we see

$$w_1(S_0) = \{\mathsf{x}: -1 \leq x_1 \leq 1 \text{ and } -1 \leq x_2 \leq 1\} \text{ and}$$

$$w_2(S_0) = \{\mathsf{x}: -\tfrac{1}{2} \leq x_1 \leq \tfrac{3}{2} \text{ and } -1 \leq x_2 \leq 1\}$$

since $w_2(S_0)$ is simply $w_1(S_0)$ translated one-half unit to the right. Therefore,

$$S_1 = W(S_0) = \{\mathsf{x}: -1 \leq x_1 \leq \tfrac{3}{2} \text{ and } -1 \leq x_2 \leq 1\}.$$

Now,

$$w_1(S_1) = \{\mathsf{x}: -\tfrac{1}{2} \leq x_1 \leq \tfrac{3}{4} \text{ and } -\tfrac{1}{2} \leq x_2 \leq \tfrac{1}{2}\} \text{ and}$$

$$w_2(S_1) = \{\mathrm{x}\colon 0 \le x_1 \le \tfrac{5}{4} \text{ and } -\tfrac{1}{2} \le x_2 \le \tfrac{1}{2}\}.$$

Thus,

$$S_2 = w_1(S_1) \cup w_2(S_1) = \{\mathrm{x}\colon -\tfrac{1}{2} \le x_1 \le \tfrac{5}{4} \text{ and } -\tfrac{1}{2} \le x_2 \le \tfrac{1}{2}\}.$$

If we keep this up, we will generate a sequence of sets $S_1, S_2, S_3, \ldots$ with

$$S_n = \{\mathrm{x}\colon -2^{-n-1} \le x_1 \le 1 + 2^{-n-2},\ -2^{-n} \le x_2 \le 2^{-n}\}.$$

So, as $n \to \infty$ S_n approaches the interval [0,1].

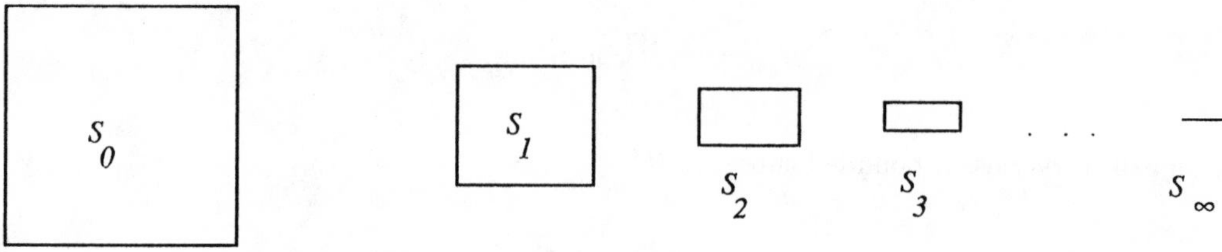

S_0 S_1 S_2 S_3 . . . S_∞

Figure 1. □

Example 2. Consider the IFS consisting of the two functions of Example 1 and the function $w_3(\mathrm{x}) = \frac{1}{2}(\mathrm{x} - (1,1)) + (1,1) = \frac{1}{2}\mathrm{x} + (\frac{1}{2},\frac{1}{2})$. We start with the triangle whose vertices are the fixed points of w_1, w_2, w_3 ((0,0), (1,0), (1,1) respectively):

$$S_0 = \{\mathrm{x}\colon 0 \le x_1 \le 1 \text{ and } 0 \le x_2 \le x_1\}.$$

We examine the image of S_0 under w_1 in detail. The vertex (0,0) is kept fixed by w_1; the vertex (1,1) is taken to $(\frac{1}{2},\frac{1}{2})$ and the vertex (1,0) is taken to $(\frac{1}{2},0)$. So, $w_1(S_0)$ is the triangle with vertices (0,0), $(\frac{1}{2},\frac{1}{2})$, $(\frac{1}{2},0)$. Similarly, the images of S_0 under w_2 and w_3 are triangles inside of S_0, see Figure 2.

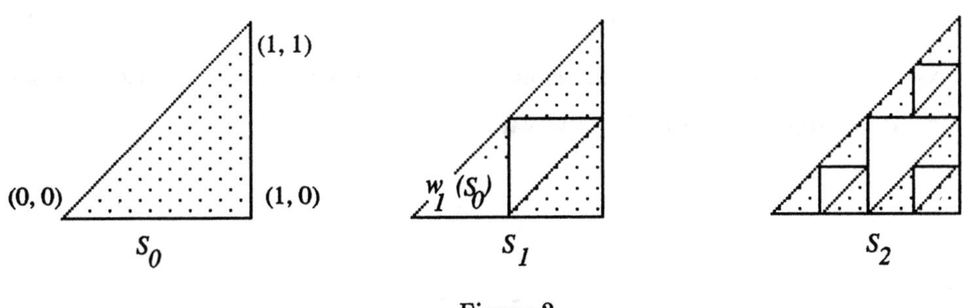

Figure 2.

Now, since $S_1 \subseteq S_0$ and since the w_i's are affine, we can sketch $S_2 = W(S_1) = w_1(S_1) \cup w_2(S_1) \cup w_3(S_1)$ simply by drawing scaled down versions of S_1 in the corners occupied by the $w_i(S_0)$'s. The effect of this is to remove an inverted right triangle from each $w_i(S_0)$. S_3 is attained by replacing the $w_i(S_1)$'s by $w_i(S_2)$'s. In the limit, after having removed all of the inverted right triangles, we obtain S_∞ which is called a *Sierpinski gasket*. □

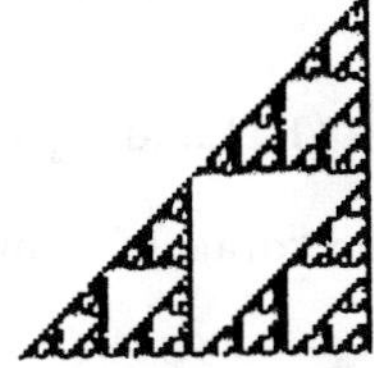

Figure 3.

Random Iteration

Another way of defining an iteration scheme for the IFS $\{w_1,...,w_d\}$ is to start with a point x_0 and select an integer between 1 and n at random (for example, if $d = 6$, we could throw a die; in a computer implementation, we would use a random number generator). If the integer k is chosen, define $\mathsf{x}_1 = w_k(\mathsf{x}_0)$. In general, if we have generated $\mathsf{x}_0,\mathsf{x}_1,...,\mathsf{x}_{n-1}$ we define $\mathsf{x}_n = w_k(\mathsf{x}_{n-1})$ if k was chosen at random from $\{1,...,d\}$. If the IFS is sufficiently well behaved, the x_n's generated this way will eventually cluster around a distinguished set which does not depend on x_0. This set, called the *attractor* of the IFS, is the counterpart of the fixed point of a single affine function and is in fact a fixed set of the associated set-valued function W.

Example 3. Let w_1 and w_2 be as in Example 1. Take $\mathsf{x}_0 = (5,5)$. We generated 1050 points according to the scheme $\mathbf{x}_n = w_i(\mathbf{x}_{n-1})$ where i was chosen at random by a computer. The last 1000 points were plotted in Figure 4 and appear to fill up the interval [0,1].

Figure 4. The Attractor of the IFS of Example 1. □

The affine functions in the examples were all of the form (1). If we allow rotations in addition to scalings the attractors become more interesting.

Example 4. With $w_1(\mathbf{x}) = \frac{1}{\sqrt{2}} R_{\frac{\pi}{4}}(\mathbf{x} - (1,1)) + (1,1)$ and $w_2(\mathbf{x}) = \frac{1}{\sqrt{2}} R_{\frac{\pi}{4}}(\mathbf{x} + (1,1)) - (1,1)$ where $R_{\frac{\pi}{4}}$ is the counterclockwise by $\frac{\pi}{4}$, we obtained the "dragon" shown in Figure 5 by plotting the last 10000 iterates of 10050 iterates.

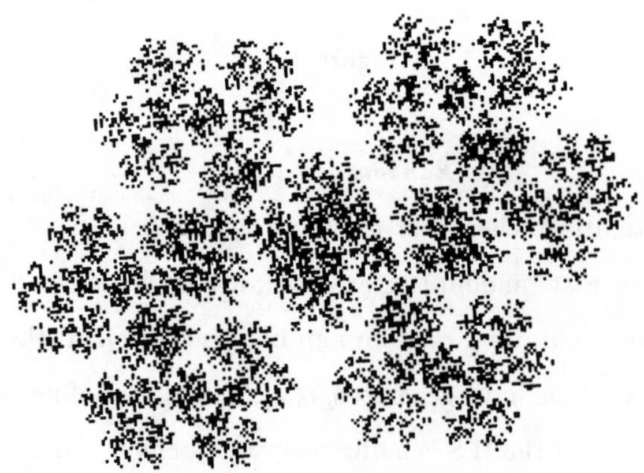

Figure 5. Dragon Attractor.

Theorem 2. Let $w_1,\ldots,w_d$ be an IFS with each w_i of the form

$$w_i(\mathsf{x}) = \alpha_i R_{\theta_i}\mathsf{x} + \mathsf{c}_i$$

where c_i is in $\mathbb{R}^2$, R_{θ_i} is the 2×2 rotation matrix of θ_i radians and $0 < \alpha_i < 1$. Then, there is a unique closed, bounded set $\mathcal{A}$ of $\mathbb{R}^2$ (the attractor) so that

(a) For any initial x_0, the random iteration $\mathsf{x}_n = w_i(\mathsf{x}_{n-1})$, i chosen at random from $\{1,\ldots,d\}$, clusters about $\mathcal{A}$.

(b) For any initial closed, bounded set S_0, the iteration $S_n = W(S_{n-1})$ converges to $\mathcal{A}$.

(c) $\mathcal{A} = w_1(\mathcal{A}) \cup w_2(\mathcal{A}) \cup \cdots \cup w_n(\mathcal{A}) = W(\mathcal{A})$.

Property (c) characterizes $\mathcal{A}$ as the *fixed-set* of the set-valued function W. Given a subset T of the place one can try to find an IFS whose attractor is T. According to (c) we should look for affine functions that take T onto subsets of itself with the images of T covering T.

For example, with $w_1(\mathsf{x}) = \frac{1}{2}\mathsf{x}$ and $w_2(\mathsf{x}) = \frac{1}{2}\mathsf{x} + (\frac{1}{2},0)$, we see that

$$w_1([0,1]) = [0,\tfrac{1}{2}] \text{ and } w_2([0,1]) = [\tfrac{1}{2},1].$$

So,

$$[0,1] = w_1([0,1]) \cup w_2([0,1]).$$

Example 5. (A true story.) A researcher picked some ivy leaves of various sizes on his way to work. He covered the largest leaf with four of the smaller leaves and estimated that the portion of the large leaf occupied by each of the smaller leaves could be obtained by scalings and rotations of the large leaf. After some fine tuning the following functions were defined:

$$w_1(\mathsf{x}) = \tfrac{1}{2}R_{\frac{\pi}{4}}(\mathsf{x} + (0.4,\ 0.2)) - (0.4,\ 0.2)$$

$$w_2(\mathsf{x}) = \tfrac{1}{2}R_{-\frac{\pi}{4}}(\mathsf{x} - (0.4,\ -0.2)) + (0.4,\ -0.2)$$

$$w_3(\mathsf{x}) = \tfrac{1}{2}(\mathsf{x} - (0,\ 0.8)) + (0,\ 0.8)$$

$$w_4(\mathbf{x}) = \tfrac{1}{2}(\mathbf{x} - (0,\ -0.3)) - (0,\ -0.3).$$

Here is the attractor generated by the random iteration method:

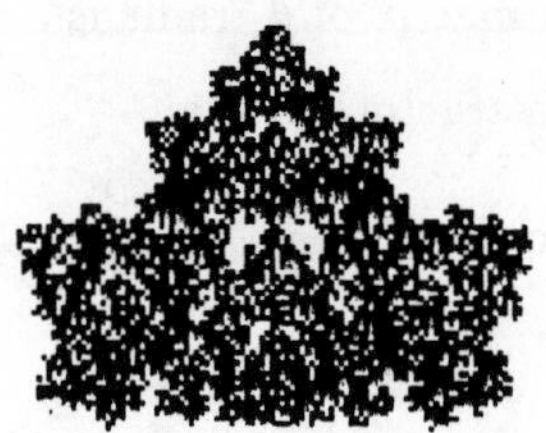

Figure 6. Ivy Leaf Attractor. □

For further information concerning the IFS method of fractal generation, see the article "A better way to compress images" by Barnsley and Sloan in *Byte* magazine, January, 1988.

EXERCISES

1. Modify the IFS of Example 2 to consist of functions of the form $w_i(\mathbf{x}) = 0.5R_\theta(\mathbf{x}-\mathbf{p})_i + \mathbf{p}_i$ and plot a few thousand of the points thus generated.

2. Let $\mathbf{p}_1,\mathbf{p}_2,\ldots,\mathbf{p}_k$ be points in the plane. Let $0 < \alpha < 1$ be a fixed number and define $w_i(\mathbf{x}) = \alpha(\mathbf{x}-\mathbf{p}_i) + \mathbf{p}_i$ for $i = 1,2,\ldots,k$. It then follows that the attractor of the IFS $\{w_1,\ldots,w_k\}$ is a subset of the polygon whose vertices are $\mathbf{p}_1,\ldots,\mathbf{p}_k$. For a specific choice of five vertices, e.g., the vertices of the pentagon P used in Chapter 4, generate the attractors that correspond to different choices of α, say $\alpha = 0.1,\ 0.4,\ 0.6,\ 0.9$.

3. Modify the IFS of Problem 1 to consist of maps of the form $w_i(\mathbf{x}) = \alpha R_\theta(\mathbf{x}-\mathbf{p}_i) + \mathbf{p}_i$ for $i = 1,2,\ldots,5$ and generate attractors corresponding to different values of α and θ. The attractor is not necessarily a subset of the pentagon determined by $\mathbf{p}_1,\ldots,\ \mathbf{p}_5$. Can you determine a larger polygon which must contain the attractor?

Comments on Selected Exercises

Section 1.2. Exercise 5. The unique solution of the system is $x = 2$, $y = -2$. If you round-off your computations, you will not get this answer. This is an example of an *ill-conditioned* system of equations due to W. Kahan.

Section 2.1. Exercise 5. 0^t is the vector [0 0].

Exercise 7. $x^tAx = 2x^2 + 2xy + 2y^2 = x^2 + 2xy + y^2 + x^2 + y^2 = (x+y)^2 + x^2 + y^2 \geq 0.$

Exercise 8. $x^tBx = x^2 + 2xy + y^2 = (x+y)^2 \geq 0$ for all x,y.

Exercise 9. Let $A = \begin{bmatrix} \mathbf{a}_1^t \\ \mathbf{a}_2^t \end{bmatrix}$ and $B = [\mathbf{b}_1 \quad \mathbf{b}_2]$. Then,

$$AB = \begin{bmatrix} \mathbf{a}_1^t\mathbf{b}_1 & \mathbf{a}_1^t\mathbf{b}_2 \\ \mathbf{a}_2^t\mathbf{b}_1 & \mathbf{a}_2^t\mathbf{b}_2 \end{bmatrix}$$

and

$$B^tA^t = \begin{bmatrix} \mathbf{b}_1^t \\ \mathbf{b}_2^t \end{bmatrix} [\mathbf{a}_1 \quad \mathbf{a}_2] = \begin{bmatrix} \mathbf{b}_1^t\mathbf{a}_1 & \mathbf{b}_1^t\mathbf{a}_2 \\ \mathbf{b}_2^t\mathbf{a}_1 & \mathbf{b}_2^t\mathbf{a}_2 \end{bmatrix}$$

Section 2.2. Exercise 4. $MM = R^tSRR^tSR = R^tSSR$ since $RR^t = I$. Eventually, $MM = I$.

Section 2.3. Exercise 3. Since $(1.2969)(0.1441) - (0.2161)(0.8648) = 10^{-8}$, the entries of the inverse are quite large.

Exercise 4. Matrices having the property defined in (a) are called *orthogonal matrices.*

Section 3.1. Exercise 7. This is an example of a *total least squares problem.* A general treatment is beyond the scope of this book.

Section 3.2. Exercise 8 indicates one potential problem with the normal equations: if the columns of A are just barely linearly independent, then the columns of A^tA could lose linear independence in computer arithmetic. In practice, if one suspected this phenomenon, one could use an alternative method called the QR factorization.

Exercise 9. AA^t is the matrix that takes $\mathbf{b}$ to $\mathbf{b}^*$; that is, it is the matrix that projects onto the column space of A.

Section 3.3. Exercise 6. In Exercise 1(f), the null space equals the column space.

Exercise 7. No, the sum of the dimensions is an odd integer.

Section 4.3. Exercise 5. The given circle is tangent to the sides of the given square. Its image is the ellipse tangent to the sides of the rectangle.

Section 4.4. Exercise 8. $\begin{bmatrix} \cos\theta & -\sin\theta \\ \sin\theta & \cos\theta \end{bmatrix} = \begin{bmatrix} \cos\theta & \sin\theta \\ \sin\theta & -\cos\theta \end{bmatrix}\begin{bmatrix} 1 & 0 \\ 0 & -1 \end{bmatrix}$ tells it all.

Section 5.4. Exercise 5. If a 2×2 matrix has distinct real eigenvalues, then the system $\mathbf{x}' = A\mathbf{x}$ is solved by using $A = UDU^{-1}$ and solving the trivial system $\mathbf{u}' = D\mathbf{u}$. If there is only one eigenvalue, then this approach fails. This exercise suggests a treatment of this case. We can find an upper triangular matrix T and an invertible matrix U so that $A = UTU^{-1}$. Then, the solution of $\mathbf{x}' = A\mathbf{x}$ is equivalent to the solution of the simple (but not trivial system) $\mathbf{u}' = T\mathbf{u}$.

Appendix B. Exercise 3. With $\mathbf{x}_0$ you go to (3.4874, 2.2616); with $\mathbf{z}_0$ to (1.459, −1.397).

Exercise 4. Since the coefficient of z^2 is 0, the sum of the roots is 0. The polynomial $z^3 - z - 1$ changes sign in [1,2]. This information should aid in determining initial guesses.

Exercise 5. The cover contains two pictures made this way, one for $-1 \leq x \leq 1$, $-1 \leq y \leq 1$ and one for a subset of this window.

Appendix D. Examples of Exercises 1 and 2 appear on the cover.

INDEX